Subversion of immune cell signalling by parasites

EDITED BY WILLIAM HARNETT

CO-ORDINATING EDITOR L. H. CHAPPELL

Preface

One of the most studied fields within the discipline of parasitology in the 20th century was immuno-parasitology. Although the work undertaken extended into areas such as immunodiagnosis, the main emphasis was on understanding the interaction between the host immune system and the invading parasite in order that information could be gained that would aid parasite control and hence improve human health. Ultimately, the goal was to produce safe and effective vaccines against diseases such as malaria and schistosomiasis. This has proved much more difficult than originally anticipated but a wealth of data was generated some of which can help us understand the difficulties associated with vaccine development. For example, we came to learn of parasite immune system evasion strategies, including fascinating mechanisms such as acquisition of host antigens in schistosomes and antigenic variation in trypanosomes. However, an alternative approach to successful parasitism, rather than to avoid the host immune system, is to actually target it for manipulation. Eukaryotic parasites have proved equally adept at this.

A strategy facilitating survival that has been described for virtually every form of eukaryotic parasite in which it has been sought is the induction of some form of 'immunosuppression'. This term tends to be used by immunoparasitologists when some form of impaired immune response has been witnessed. Its possible existence has been remarked on since the beginning of the 20th century when it was observed that people infected with eukaryotic parasites often appeared more susceptible to bacterial infections or were more difficult to vaccinate. Furthermore, numerous studies, some going back many years, have shown that animals infected with eukaryotic parasites, demonstrate impaired immune responses to heterologous antigens. Analysis of the mechanisms underlying these defects has revealed that a number

of effector mechanisms of the immune system are targeted and indeed the identity of particular cells that are affected has been reported.

This supplement is specifically concerned with the actions parasites must take to control the activities of cells of the host immune system. However, whereas in the past research has been concerned with measuring the effect of parasites on immunological parameters, advances in biochemistry and cell biology in the last two decades have permitted a molecular dissection of the functionalism of cells. The main focus here is thus on how parasites have evolved strategies to subvert the pathways that immune system cells employ to transduce signals pertinent to parasite elimination.

The six articles that the supplement contains were produced by speakers (and their colleagues) at the 2004 British Society for Parasitology Autumn Symposium on 'Subversion of immune cell signalling by parasites'. The first article by Goodridge and Harnett (M.) represents an introduction to the topic by familiarising readers with the principles of intracellular signal transduction and outlining the major immune signalling pathways triggered following the ligation of antigen receptors, cytokine receptors and Toll-like receptors (TLRs). It thus gives an overview of the cellular targets for subversion by parasites. The next article by Haga and Bowie is concerned with subversion of immune cell signalling in a very well defined system, *Vaccinia* virus. By taking note of what has been learned in such a system, parasitologists may acquire ideas of relevance to studying their own particular parasites. The remainder of the chapters focus on immune subversion mechanisms employed by eukaryotic parasites. Gregory and Olivier write on inhibition of macrophage signal transduction, describing a role for the host tyrosine phosphatase SHP-1 in the induction of macrophage

Parasitology (2005), **130**, S1–S2. © 2005 Cambridge University Press
doi:10.1017/S0031182005008334 Printed in the United Kingdom

dysfunction by *Leishmania* parasites. Langsley and colleagues discuss the reversible transformation of host lymphocytes by *Theileria* parasites and the roles that casein kinase II may play in this. Schofield and colleagues focus on the malarial GPI toxin and its interaction with host receptors. Structural requirements for interaction are discussed and the role of TLR-2 as a receptor highlighted. In the last article, Harnett (W.) and colleagues describe how the filarial nematode-secreted product ES-62 modulates the activity of various cells of the immune system by selectively targeting signal transduction pathways involved in activation, proliferation and polarisation.

Finally, I would like to thank the following people: all the presenters at the symposium and/or authors who agreed to write the articles; Les Chappell in his role as co-ordinator for the supplement; the accommodating staff at the Linnean Society where the meeting was held, members of the Harnett labs for their assistance on the day and each of Active Motif, Cambridge University Press, Miltenyi Biotec, Pfizer Ltd. and 'Trends in Parasitology' for financial support.

W. HARNETT
May 2005

CAMBRIDGE
UNIVERSITY PRESS

Shaftesbury Road, Cambridge CB2 8EA, United Kingdom

One Liberty Plaza, 20th Floor, New York, NY 10006, USA

477 Williamstown Road, Port Melbourne, VIC 3207, Australia

314–321, 3rd Floor, Plot 3, Splendor Forum, Jasola District Centre, New Delhi – 110025, India

103 Penang Road, #05–06/07, Visioncrest Commercial, Singapore 238467

Cambridge University Press is part of Cambridge University Press & Assessment, a department of the University of Cambridge.

We share the University's mission to contribute to society through the pursuit of education, learning and research at the highest international levels of excellence.

www.cambridge.org
Information on this title: www.cambridge.org/9780521684866

DOI:10.1017/S0031182005008115

© Cambridge University Press & Assessment 2005

A catalogue record for this publication is available from the British Library

ISBN   978-0-521-68486-6   Paperback

Cambridge University Press & Assessment has no responsibility for the persistence or accuracy of URLs for external or third-party internet websites referred to in this publication and does not guarantee that any content on such websites is, or will remain, accurate or appropriate.

Subscriptions may be sent to any bookseller or subscription agent or direct to the publisher : Cambridge University Press, The Edinburgh Building, Shaftesbury Road, Cambridge CB2 2RU, UK. Subscriptions in the USA, Canada and Mexico should be sent to Cambridge University Press, Journals Fulfillment Department, 100 Brook Hill Drive, West Nyack, New York 10994Ð2133. All orders must be accompanied by payment. The subscription price (excluding VAT) of volumes 130 and 131, 2005, which includes print and electronic access, is £620 (US $1022 in the USA, Canada and Mexico), payable in advance, for twelve parts plus supplements; the electronic-only price for institutional subscribers is £525 (US $855 in the USA, Canada and Mexico); separate parts cost £51 or US $82 each (plus postage). EU subscribers (outside the UK) who are not registered for VAT should add VAT at their countryÒs rate. VAT registered subscribers should provide their VAT registration number. Japanese prices for institutions are available from Kinokuniya Company Ltd, P.O. Box 55, Chitose, Tokyo 156, Japan. Prices include delivery by air. Periodicals postage paid at New York, NY and at additional mailing ofÞces. POSTMASTER : send address changes in USA, Canada and Mexico to Parasitology, Cambridge University Press, 100 Brook Hill Drive, West Nyack, New York 10994Ð2133.

Introduction to immune cell signalling

H. S. GOODRIDGE *and* M. M. HARNETT

Division of Immunology, Infection and Inflammation, University of Glasgow, Glasgow, G11 6NT, UK

SUMMARY

The dynamic interaction of cells of the immune system with other cells, antigens and secreted factors determines the nature of an immune response. The response of individual cells is governed by the sequence of intracellular signalling events triggered following the association of cell surface molecules during cell-cell contact or the detection of soluble molecules of host or pathogen origin. In this review we will first outline the general principles of intracellular signal transduction. We will then describe the signalling pathways triggered following the recognition of antigen, as well as the detection of cytokines, and discuss how the signalling pathways activated regulate the effector response.

Key words: signal transduction, BCR signalling, cytokine receptor signalling, TLR signalling.

BASIC PRINCIPLES OF CELL SIGNALLING

Cells sense, respond to and integrate a multiplicity of signals from their environment. In the context of the immune system, these include signals resulting from interactions with neighbouring cells (cell-cell contact), and the detection of soluble factors such as cytokines, which may originate at distant sites and be transported via the blood, as well as the recognition of pathogens and their products. Following ligation of cell surface receptors, intracellular signal transduction cascades are initiated, resulting in the activation of transcription factors and other proteins that regulate processes such as gene induction, phagocytosis, apoptosis, proliferation and secretion (Harnett, 1999 *a, b*).

Signal transduction cassettes comprise specific cell-surface membrane receptors, effector signalling elements and regulatory proteins. These signalling cassettes serve to detect, amplify and integrate diverse external signals to generate the appropriate cellular response. In this review, we first discuss how cell-surface receptors sense and transduce signals by transmembrane coupling to effector enzyme systems, generating low-molecular-weight molecules termed second messengers and activating a range of key protein kinases with distinct substrates that control gene induction and other cellular responses (Harnett, 1999 *a, b*). We will then describe the signalling cascades underlying some of the key responses of cells of the immune system to environmental stimuli of host as well as pathogen origin.

Phosphorylation is an important mechanism of control of protein recruitment and activation; kinases catalyse the transfer of a phosphate group from ATP onto specific amino acids, a process that can be reversed by the action of phosphatases. Individual kinases have distinct substrate specificities e.g. tyrosine kinases, dual specificity serine/threonine or tyrosine/threonine kinases (Bromann, Korkaya & Courtneidge, 2004; Cannons & Schwartzberg, 2004; Gambacorti-Passerini, 2004; Piiper & Zeuzem, 2004). Lipid kinases, such as phosphoinositide-3-kinase (PI3K), catalyse the phosphorylation of lipids (Cantrell, 2001, 2002, 2003 *b*).

Recruitment of kinases and other enzymes, such as phospholipases, to the activated receptor following ligation triggers a cascade of enzymatic activation and the generation of second messengers such as cyclic AMP (cAMP) (Houslay, 1998; Houslay & Adams, 2003; Houslay & Baillie, 2003), inositol trisphosphate (IP3) (Irvine, 2003 *a, b*; 2004), diacylglycerol (DAG) (Powner & Wakelam, 2002; McDermott, Wakelam & Morris, 2004) and calcium ions (Ca^{2+}) (Berridge, Bootman & Roderick, 2003; Berridge, 2004), which results in propagation and amplification of the signal. Hydrolysis of phosphatidylcholine, which comprises about 40% of the total cellular phospholipid, by various phospholipases results in the generation of a range of second messengers, including arachidonic acid (phospholipase A2, PLA2), DAG (phospholipase C, PLC) and phosphatidic acid (phospholipase D, PLD), depending on the site of action of the enzyme (Wakelam & Harnett, 1998; Powner & Wakelam, 2002; McDermott *et al.* 2004). Arachidonic acid is a key lipid second messenger involved in the regulation of signalling enzymes, such as PLC-γ and -δ, and the α, β and γ isoforms of the protein kinase C (PKC) family of serine/threonine kinases. DAG is a cofactor required for the activation of PKC isoforms; it can

Address for correspondence: Prof. M. M. Harnett, Division of Immunology, Infection and Inflammation, University of Glasgow, Western Infirmary, Dumbarton Road, Glasgow G11 6NT, U.K. Tel: 0141-211-2247, Fax: 0141-337-3217. E-mail: M.Harnett@bio.gla.ac.uk

Parasitology (2005), **130**, S3–S9. © 2005 Cambridge University Press
doi:10.1017/S0031182005008115   Printed in the United Kingdom

also be metabolised to generate arachidonic acid by the action of DAG lipase. Furthermore, phosphatidic acid can be interconverted to DAG by the action of phosphatidic acid phosphohydrolase (Wakelam & Harnett, 1998; Powner & Wakelam, 2002; McDermott *et al.* 2004).

One of the key outcomes of intracellular signalling cascades is the activation of transcription factors, such as NF-κB, NFAT, Fos, Jun and Oct. Transcription requires the binding of RNA polymerase to the promoter region situated 5′ of the transcription start site. Transcription factors, which bind to specific regulatory DNA sequences, co-operate to permit or deny RNA polymerase access to the promoter, and thereby regulate gene expression (Clevenger, 2004; Coffer & Burgering, 2004; Eggert *et al.* 2004; Smith & Sigvardsson, 2004).

ANTIGEN RECEPTOR SIGNALLING

Recognition of specific antigens by B cells is achieved by surface-bound immunoglobulin (sIg). However, sIg is unable to transduce intracellular signals due to its short cytoplasmic tail, and hence recruits two Ig-α/β heterodimers, which possess Intracellular Tyrosine-based Activatory Motifs (ITAMs; D/E-X_7-D/E-XX-YXXL-X_7-YXXL/I) in their cytoplasmic tails. The earliest detectable B cell receptor (BCR) signal is tyrosine phosphorylation of proteins including the Ig-α/β ITAMs; mutations in either of the conserved tyrosines disrupts signalling. ITAM phosphorylation results in the initiation of a signalling scaffold around the active BCR (see Fig. 1). The tyrosine kinases Lyn and Syk bind to the phosphorylated ITAMs via their SH2 domains (Dal Porto *et al.* 2004; Gauld & Cambier, 2004; Harnett, Katz & Ford, 2005). Further recruitment of adaptors/signalling proteins occurs via SH2, SH3 (binds proline-rich sequences) and PH (binds phosphoinositides) domain interactions. The tyrosine phosphatase SHP-1 also associates with the BCR to prevent aberrant activation (Kurosaki, 2002; Cannons & Schwartzberg, 2004; Simeoni *et al.* 2004).

The phosphoinositide-3-kinase (PI3K) pathway is a key pathway promoting survival, gene induction and cell cycle progression following B cell activation (Fruman, 2004*a*,*b*). PI3K catalyses the conversion of phosphatidylinositol-4,5-bisphosphate (PIP2) to phosphatidylinositol-3,4,5-trisphosphate (PIP3). This enables the recruitment and activation of the serine/threonine kinase protein kinase B (PKB), also known as Akt, which prevents apoptosis and promotes B cell survival, as well as the serine/threonine kinase p70 S6 kinase, which is required for the entry of cells into S phase of the cell cycle. Akt phosphorylates the pro-apoptotic protein Bad, which is then sequestered by association with the 14-3-3 protein, thereby preventing the release of cyto-

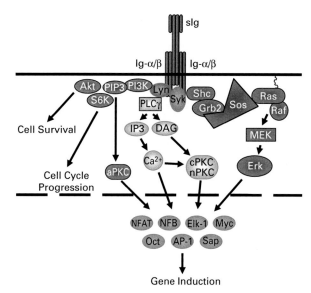

Fig. 1. B cell receptor (BCR) signalling. Surface-bound immunoglobulin (sIg) alone cannot transduce intracellular signals following recognition of specific antigen. Instead signalling is initiated by tyrosine phosphorylation of Ig-α/β dimers, which associate with sIg to form the BCR complex. A signalling scaffold forms around the activated receptor complex with recruitment of adaptors and other signalling proteins, which trigger the activation of various cascades including the PI3K, PLC-γ and MAP kinase pathways, ultimately resulting in the promotion of gene induction, B cell survival and proliferation.

chrome c and activation of the caspase cascade. Thus PIP3 promotes B cell survival and proliferation (Cantrell, 2001, 2002, 2003*b*; Fruman, 2004*a*).

PIP3 can also regulate gene induction via the activation of atypical isoforms of PKC (ε and ζ), which activate transcription factors such as AP-1, NFAT and NF-κB (Hao, Kurosaki & August, 2003; Guo, Su & Rawlings, 2004). Furthermore, PIP3 can recruit PLC-γ, which hydrolyses PIP2 to generate IP3 and DAG, resulting in the release of Ca^{2+} ions from intracellular stores and the activation of classical and novel isoforms of PKC. Ca^{2+} ion binding is required for the serine/threonine kinase activity of the Ca^{2+}/calmodulin (CaM)-dependent protein kinase (CaMKII) and enables the CaM protein phosphatase calcineurin to dephosphorylate and activate NFAT. DAG is necessary for the activation of various PKC isoforms, which may also require Ca^{2+} binding, resulting in the activation of transcription factors such as AP-1, NFAT, NF-κB and Oct (Hao *et al.* 2003; Guo *et al.* 2004).

Another key event following ligation of the BCR is the activation of the mitogen-activated protein kinase (MAP kinase) pathways. There are three major subfamilies of MAP kinases – the extracellular regulated kinases (Erk1/2), the p38 MAP kinases and the c-Jun kinases/stress-activated protein kinases (JNK/SAPK) (Adams *et al.* 2004; Roux & Blenis,

2004; Rubinfeld & Seger, 2004; Harnett *et al.* 2005). Recruitment of a GTPase to the activated receptor complex is required for the activation of a kinase cascade, ultimately leading to activation of MAP kinases, which target transcription factors to regulate gene induction. For example, BCR coupling to the Erk cascade is achieved via the GTPase Ras (see Fig. 1). Inactive Ras (Ras-GDP) is anchored at the plasma membrane and is recruited by the Grb2-Sos complex. Sos, a guanine nucleotide exchange factor (GEF), promotes the exchange of GDP for GTP. Active Ras (Ras-GTP) activates Raf (a MAP kinase kinase kinase) by binding to its N-terminal regulatory domain. Raf then phosphorylates MEK-1 (a MAP kinase kinase), which in turn phosphorylates Erk1/2. The Erk MAP kinase pathway promotes the proliferation and differentiation of B cells by targeting transcription factors such as c-Myc, Sap and Elk-1 (Cantrell, 2003*a, b*, Dal Porto *et al.* 2004; Harnett *et al.* 2005).

CYTOKINE RECEPTOR SIGNALLING

Cytokines and their receptors can be classified according to their structure. Most cytokine receptors belong to the class I cytokine receptor family; they are multichain receptor complexes comprising separate chains for specific cytokine recognition and signal transduction. Signalling chains may be shared by several cytokines within a subfamily. For example, IL-3, IL-5 and GM-CSF, which are structurally similar, share a common β chain, while IL-2, IL-4, IL-7, IL-9 and IL-15 share a common γ chain. gp130 is the signalling subunit of the IL-6 and IL-11 receptors, and the gp130-related IL-12Rβ1 chain is utilised by IL-12 and IL-23 (Hofmann *et al.* 2002; O'Shea, 2004; O'Shea *et al.* 2004).

Most cytokines signal through PLC, PI3K, and Ras MAPK pathways, as well as JAK-STAT pathways (reviewed in Leonard & Lin, 2000; Hofmann *et al.* 2002; Agnello *et al.* 2003; O'Shea, 2004; O'Shea *et al.* 2004). Binding of cytokines to their receptors triggers the recruitment of receptor-associated tyrosine kinases of the Janus kinase (JAK) family (JAK1-3, Tyk2). Activated JAKs phosphorylate tyrosine residues in the β and γ signalling receptor chains, enabling the recruitment of other proteins via their SH2 domains, in particular the cytosolic STAT proteins (signal transducers and activators of transcription; STAT1-6). STATs are then phosphorylated and released from the receptor. Phosphorylated STAT proteins homodimerise or heterodimerise via pTyr-SH2 domain interactions, and rapidly translocate into the nucleus where they act as transcription factors by binding to GAS motifs (see Fig. 2).

Different combinations of JAKs and STATs can be utilised to achieve specificity. For example, the JAK-STAT pathways play key roles in the control of

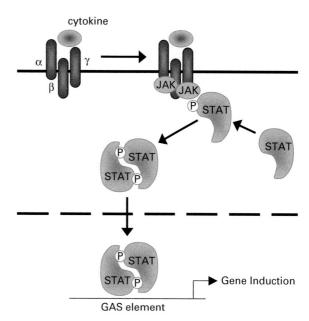

Fig. 2. Cytokine receptor signalling via JAK-STAT pathways. Following cytokine receptor ligation, JAKs are recruited and phosphorylate tyrosine residues in the β and γ signalling receptor chains, enabling STAT recruitment and phosphorylation. pSTAT proteins form dimers, which rapidly translocate to the nucleus where they act as transcription factors by binding to GAS elements.

polarisation of CD4$^+$ helper T (Th) cell responses (Murphy *et al.* 1999, 2000; Murphy & Reiner, 2002; Agnello *et al.* 2003). Cytokine signals received by Th precursors determine their differentiation towards a Th1 or Th2 phenotype. Antigen recognition by the TCR in the presence of IL-12 and IFN-γ leads to the development of Th1 responses, while a Th2 bias is achieved by exposure to IL-4. Polarisation is achieved via the action of two transcription factors, T-bet and GATA-3 (Zheng & Flavell, 1997; Szabo *et al.* 2000; Murphy & Reiner, 2002; see Fig. 3). IL-12 and IFN-γ signal through their receptors on the surface of precursor Th cells, leading to the activation of STAT4 and STAT1 respectively, both of which induce T-bet expression. T-bet promotes IFN-γ expression by chromatin remodelling and increases IL-12Rβ2 chain expression, thereby further enhancing the IL-12 and IFN-γ signals, in part by sustaining its own expression.

Conversely, recognition of IL-4 by its receptor triggers the activation and nuclear translocation of STAT6, which rapidly induces GATA-3 expression (Zheng & Flavell, 1997; Zhou & Ouyang, 2003). GATA-3 regulates the expression of Th2 cytokines by co-ordinate regulation of the Th2 locus, which contains the genes encoding IL-4, IL-5 and IL-13, by not only inducing the expression of IL-5 and IL-13, but also by promoting chromatin remodelling to enable IL-4 transcription (Murphy *et al.* 1999, 2000; Lee *et al.* 2000; Ouyang *et al.* 2000; Murphy & Reiner, 2002; Agnello *et al.* 2003). GATA-3 thereby

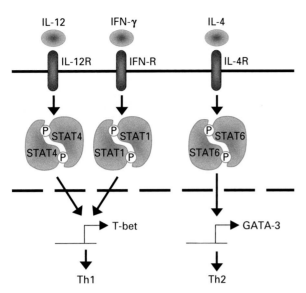

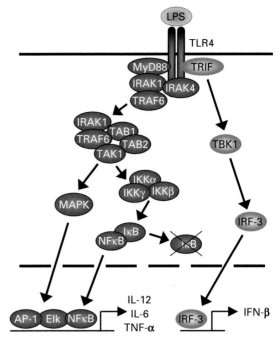

Fig. 3. Signalling pathways underlying Th1/Th2 polarisation. JAK-STAT pathways underlie the polarisation of CD4[+] helper T (Th) cell responses. IL-12 and IFN-γ signal via STAT4 and STAT1 respectively to induce expression of the transcription factor T-bet, which promotes Th1 polarisation via enhanced IFN-γ and IL-12Rβ2 expression. IL-4 induces activation of STAT6, leading to production of GATA-3, which promotes transcription of the Th2 cytokines IL-4, IL-5 and IL-13. Positive and negative feedback mechanisms reinforce the Th1/Th2 bias.

Fig. 4. Toll-like receptor 4 (TLR4) signalling. Following LPS binding to its receptor complex, signalling is initiated via TLR4. Adaptor proteins, including MyD88 and TRIF, are recruited, as well as IRAK-1, IRAK-4 and TRAF6. Following phosphorylation of IRAK-1 by IRAK-4, and TRAF6 by IRAK-1, IRAK-1 and TRAF6 dissociate from the receptor and form a complex with TAK1, TAB1 and TAB2. TAK1 activates the MAP kinase and NF-κB pathways, resulting in transcription of pro-inflammatory cytokines. A MyD88-independent, TRIF-dependent pathway via TBK-1 results in phosphorylation of the transcription factor IRF-3, which induces IFN-β production.

augments its own expression by positive feedback autoregulation. Furthermore, negative feedback mechanisms reinforce the fate decision. GATA-3 inhibits Th1 development by suppressing T-bet expression, at least in part by downregulating STAT4, and by increasing IL-4 production, thereby suppressing IL-12Rβ2 chain expression (reviewed in Ho & Glimcher, 2002). Similarly, T-bet may also inhibit GATA-3 expression and IL-4 production (Murphy *et al.* 1999, 2000; Murphy & Reiner, 2002; Agnello *et al.* 2003).

TOLL-LIKE RECEPTOR SIGNALLING

Detection of pathogen products by cells of the innate immune system is mediated by pattern recognition receptors (PRR), with or without the aid of soluble host factors such as complement. PRRs include scavenger receptors, complement receptors and the recently identified family of Toll-like receptors (TLRs). TLRs, which were originally identified by their homology to the *Drosophila* anti-fungal protein Toll, have been demonstrated to be crucial for the recognition of a variety of pathogens and their products, including proteins, carbohydrates and nucleotides derived from bacteria, viruses, fungi, protozoa and helminth parasites (reviewed in Takeda, Kaisho & Akira, 2003).

TLRs, which are members of the Toll/IL-1R (TIR) superfamily, comprise extracellular

leucine-rich repeats (LRRs), a transmembrane region and an intracellular region containing a conserved intracellular TIR domain. TLRs signal via the recruitment of adaptor proteins, such as MyD88, TIR-containing adaptor protein/MyD88 adaptor-like (TIRAP/Mal), TIR-containing adaptor inducing IFN-β/TIR-containing adaptor molecule-1 (TRIF/TICAM-1), TRIF-related adaptor molecule (TRAM) and sterile α and HEAT-Armadillo motifs (SARM) (reviewed in McGettrick & O'Neill, 2004). Specificity is achieved by homo- or heterodimerisation of TLRs and differential adaptor recruitment. Different downstream events are triggered by the association of various adaptors with the TLRs e.g. MyD88-dependent NF-κB and MAP kinase activation, and TRIF-dependent activation of the transcription factor IRF-3.

The signalling pathways downstream of TLR4, the receptor required for the recognition of and response to bacterial lipopolysaccharide (LPS), are the most thoroughly investigated so far (reviewed in Takeda & Akira, 2004; see Fig. 4). A MyD88-dependent pathway is required for LPS-triggered induction of IL-12 and pro-inflammatory cytokines such as IL-6 and TNF-α. Following LPS recognition, IL-1 receptor-associated kinase (IRAK)-1, IRAK-4 and TNF receptor-associated factor (TRAF)-6 are recruited to the receptor complex. IRAK-1 associates with MyD88 via the interaction of their death domains, and is phosphorylated by IRAK-4. Phosphorylated IRAK-1 interacts with and phosphorylates TRAF6, and they dissociate from the receptor to form a complex with TGFβ-activated kinase (TAK)1, TAK1-binding protein (TAB)1 and TAB2. Activation of TAK1 results in phosphorylation of the IKK complex, which consists of the I-κB kinases IKKα, IKKβ and NEMO/IKKγ. Phosphorylation of I-κB by the IKK complex leads to its ubiquitination and degradation, releasing the transcription factor NF-κB to enter the nucleus and promote gene induction. TAK1 is also responsible for the activation of MAP kinase pathways, leading to gene induction via activation of transcription factors such as Elk, AP-1 and ATF-2.

In contrast, knockout studies indicated that MyD88 is not required for the induction of IFN-inducible genes, such as IP-10 and GARG16, following LPS stimulation (Kawai *et al.* 2001). These genes are induced indirectly via MyD88-independent production of IFN-β, which activates STAT1 (Toshchakov *et al.* 2002). The transcription factor IRF-3, which is activated by signalling via the TRIF adaptor, regulates IFN-β production. This occurs via activation of the IKK isoform TANK-binding kinase 1 (TBK1), leading to IRF-3 phosphorylation and nuclear translocation (Fitzgerald *et al.* 2003; Sharma *et al.* 2003; Hemmi *et al.* 2004).

CONCLUSIONS

Activation of intracellular signalling cascades, which integrate and amplify signals received following receptorligation enables the controlled response of cells to multiple environmental stimuli. These highly co-ordinated processes are vital for the generation of appropriate effector responses in the immune system, to achieve effective removal of pathogens while limiting host damage. However, these communication pathways are also key targets utilised by pathogens to achieve modulation of host immune responses to promote their survival and propagation.

REFERENCES

ADAMS, C. L., GRIERSON, A. M., MOWAT, A. M., HARNETT, M. M. & GARSIDE, P. (2004). Differences in the kinetics, amplitude, and localization of ERK activation in energy and priming revealed at the level of individual primary T cells by laser scanning cytometry. *Journal of Immunology* **173**, 1579–1586.

AGNELLO, D., LANKFORD, C. S., BREAM, J., MORINOBU, A., GADINA, M., O'SHEA, J. J. & FRUCHT, D. M. (2003). Cytokines and transcription factors that regulate T helper cell differentiation: new players and new insights. *Journal of Clinical Immunology* **23**, 147–161.

BERRIDGE, M. (2004). Conformational coupling: a physiological calcium entry mechanism. *Sci STKE* **2004**, pe33.

BERRIDGE, M. J., BOOTMAN, M. D. & RODERICK, H. L. (2003). Calcium signalling: dynamics, homeostasis and remodelling. *Nature Reviews Molecular and Cell Biology* **4**, 517–529.

BROMANN, P. A., KORKAYA, H. & COURTNEIDGE, S. A. (2004). The interplay between Src family kinases and receptor tyrosine kinases. *Oncogene* **23**, 7957–7968.

CANNONS, J. L. & SCHWARTZBERG, P. L. (2004). Fine-tuning lymphocyte regulation: what's new with tyrosine kinases and phosphatases? *Current Opinion in Immunology* **16**, 296–303.

CANTRELL, D. A. (2001). Phosphoinositide 3-kinase signalling pathways. *Journal of Cell Science* **114**, 1439–1445.

CANTRELL, D. (2002). Protein kinase B (Akt) regulation and function in T lymphocytes. *Seminars in Immunology* **14**, 19–26.

CANTRELL, D. A. (2003*a*). GTPases and T cell activation. *Immunology Reviews* **192**, 122–130.

CANTRELL, D. A. (2003*b*). Regulation and function of serine kinase networks in lymphocytes. *Current Opinion in Immunology* **15**, 294–298.

CLEVENGER, C. V. (2004). Roles and regulation of stat family transcription factors in human breast cancer. *American Journal of Pathology* **165**, 1449–1460.

COFFER, P. J. & BURGERING, B. M. (2004). Forkhead-box transcription factors and their role in the immune system. *Nature Reviews Immunology* **4**, 889–899.

DAL PORTO, J. M., GAULD, S. B., MERRELL, K. T., MILLS, D., PUGH-BERNARD, A. E. & CAMBIER, J. (2004). B cell antigen receptor signaling 101. *Molecular Immunology* **41**, 599–613.

EGGERT, M., KLUTER, A., ZETTL, U. K. & NEECK, G. (2004). Transcription factors in autoimmune diseases. *Current Pharmaceutical Design* **10**, 2787–2796.

FITZGERALD, K. A., McWHIRTER, S. M., FAIA, K. L., ROWE, D. C., LATZ, E., GOLENBOCK, D. T., COYLE, A. J., LIAO, S. M. & MANIATIS, T. (2003). IKKepsilon and TBK1 are essential components of the IRF3 signaling pathway. *Nature Immunology* **4**, 491–496.

FRUMAN, D. A. (2004*a*). Phosphoinositide 3-kinase and its targets in B-cell and T-cell signaling. *Current Opinion in Immunology* **16**, 314–320.

FRUMAN, D. A. (2004*b*). Towards an understanding of isoform specificity in phosphoinositide 3-kinase signalling in lymphocytes. *Biochemical Society Transactions* **32**, 315–319.

GAMBACORTI-PASSERINI, C. (2004). Oncogenic tyrosine kinases. *Cellular and Molecular Life Sciences* **61**, 2895–2896.

GAULD, S. B. & CAMBIER, J. C. (2004). Src-family kinases in B-cell development and signaling. *Oncogene* **23**, 8001–8006.

GUO, B., SU, T. T. & RAWLINGS, D. J. (2004). Protein kinase C family functions in B-cell activation. *Current Opinion in Immunology* **16**, 367–373.

HAO, S., KUROSAKI, T. & AUGUST, A. (2003). Differential regulation of NFAT and SRF by the B cell receptor via a

PLCgamma-Ca(2+)-dependent pathway. *EMBO Journal* **22**, 4166–4177.

HARNETT, M. M. (1999*a*). Cell growth, Differentiation & Cancer. In *Medical Biochemistry* (Eds, Baynes, J. and Dominiczak, M. H.) Mosby, London.

HARNETT, M. M. (1999*b*) Membrane Receptors and Signal Transduction. In *Medical Biochemistry* (Eds, Baynes, J. and Dominiczak, M. H.) Mosby, London.

HARNETT, M. M., KATZ, E. & FORD, C. A. (2005). Differential signalling during B cell maturation. *Immunology Letters*, **98**, 33–44.

HEMMI, H., TAKEUCHI, O., SATO, S., YAMAMOTO, M., KAISHO, T., SANJO, H., KAWAI, T., HOSHINO, K., TAKEDA, K. & AKIRA, S. (2004). The roles of two IkappaB kinase-related kinases in lipopolysaccharide and double stranded RNA signaling and viral infection. *Journal of Experimental Medicine* **199**, 1641–1650.

HO, I. C. & GLIMCHER, L. H. (2002). Transcription: tantalizing times for T cells. *Cell* **109** (Suppl.) S109–120.

HOFMANN, S. R., ETTINGER, R., ZHOU, Y. J., GADINA, M., LIPSKY, P., SIEGEL, R., CANDOTTI, F. & O'SHEA, J. J. (2002). Cytokines and their role in lymphoid development, differentiation and homeostasis. *Current Opinion in Allergy and Clinical Immunology* **2**, 495–506.

HOUSLAY, M. D. (1998). Adaptation in cyclic AMP signalling processes: a central role for cyclic AMP phosphodiesterases. *Seminars in Cell and Developmental Biology* **9**, 161–167.

HOUSLAY, M. D. & ADAMS, D. R. (2003). PDE4 cAMP phosphodiesterases: modular enzymes that orchestrate signalling cross-talk, desensitization and compartmentalization. *Biochemical Journal* **370**, 1–18.

HOUSLAY, M. D. & BAILLIE, G. S. (2003). The role of ERK2 docking and phosphorylation of PDE4 cAMP phosphodiesterase isoforms in mediating cross-talk between the cAMP and ERK signalling pathways. *Biochemical Society Transactions* **31**, 1186–1190.

IRVINE, R. (2004). Inositol lipids: to PHix or not to PHix? *Current Biology* **14**, R308–310.

IRVINE, R. F. (2003*a*) 20 years of Ins(1,4,5)P3, and 40 years before. *Nature Reviews Molecular and Cell Biology* **4**, 586–590.

IRVINE, R. F. (2003*b*) Nuclear lipid signalling. *Nature Reviews Molecular and Cell Biology* **4**, 349–360.

KAWAI, T., TAKEUCHI, O., FUJITA, T., INOUE, J., MUHLRADT, P. F., SATO, S., HOSHINO, K. & AKIRA, S. (2001). Lipopolysaccharide stimulates the MyD88-independent pathway and results in activation of IFN-regulatory factor 3 and the expression of a subset of lipopolysaccharide-inducible genes. *Journal of Immunology* **167**, 5887–5894.

KUROSAKI, T. (2002). Regulation of B cell fates by BCR signaling components. *Current Opinion in Immunology* **14**, 341–347.

LEE, H. J., TAKEMOTO, N., KURATA, H., KAMOGAWA, Y., MIYATAKE, S., O'GARRA, A. & ARAI, N. (2000). GATA-3 induces T helper cell type 2 (Th2) cytokine expression and chromatin remodeling in committed Th1 cells. *Journal of Experimental Medicine* **192**, 105–115.

LEONARD, W. J. & LIN, J. X. (2000). Cytokine receptor signaling pathways. *Journal of Allergy and Clinical Immunology* **105**, 877–888.

McDERMOTT, M., WAKELAM, M. J. & MORRIS, A. J. (2004). Phospholipase D. *Biochemistry and Cell Biology* **82**, 225–253.

McGETTRICK, A. F. & O'NEILL, L. A. (2004). The expanding family of MyD88-like adaptors in Toll-like receptor signal transduction. *Molecular Immunology* **41**, 577–582.

MURPHY, K. M., OUYANG, W., FARRAR, J. D., YANG, J., RANGANATH, S., ASNAGLI, H., AFKARIAN, M. & MURPHY, T. L. (2000). Signaling and transcription in T helper development. *Annual Reviews in Immunology* **18**, 451–494.

MURPHY, K. M., OUYANG, W., RANGANATH, S. & MURPHY, T. L. (1999). Bi-stable transcriptional circuitry and GATA-3 auto-activation in Th2 commitment. *Cold Spring Harbor Symposia on Quantitative Biology* **64**, 585–588.

MURPHY, K. M. & REINER, S. L. (2002). The lineage decisions of helper T cells. *Nature Reviews Immunology* **2**, 933–944.

O'SHEA, J. J. (2004). Targeting the Jak/STAT pathway for immunosuppression. *Annals of the Rheumatic Diseases* **63** (Suppl 2), ii67–ii71.

O'SHEA, J. J., HUSA, M., LI, D., HOFMANN, S. R., WATFORD, W., ROBERTS, J. L., BUCKLEY, R. H., CHANGELIAN, P. & CANDOTTI, F. (2004). Jak3 and the pathogenesis of severe combined immunodeficiency. *Molecular Immunology* **41**, 727–737.

OUYANG, W., LOHNING, M., GAO, Z., ASSENMACHER, M., RANGANATH, S., RADBRUCH, A. & MURPHY, K. M. (2000). Stat6-independent GATA-3 autoactivation directs IL-4-independent Th2 development and commitment. *Immunity* **12**, 27–37.

PIIPER, A. & ZEUZEM, S. (2004). Receptor tyrosine kinases are signaling intermediates of G protein-coupled receptors. *Current Pharmaceutical Design* **10**, 3539–3545.

POWNER, D. J. & WAKELAM, M. J. (2002). The regulation of phospholipase D by inositol phospholipids and small GTPases. *FEBS Letters* **531**, 62–64.

ROUX, P. P. & BLENIS, J. (2004). ERK and p38 MAPK-activated protein kinases: a family of protein kinases with diverse biological functions. *Microbiology and Molecular Biology Reviews* **68**, 320–344.

RUBINFELD, H. & SEGER, R. (2004). The ERK cascade as a prototype of MAPK signaling pathways. *Methods in Molecular Biology* **250**, 1–28.

SHARMA, S., TENOEVER, B. R., GRANDVAUX, N., ZHOU, G. P., LIN, R. & HISCOTT, J. (2003). Triggering the interferon antiviral response through an IKK-related pathway. *Science* **300**, 1148–1151.

SIMEONI, L., KLICHE, S., LINDQUIST, J. & SCHRAVEN, B. (2004). Adaptors and linkers in T and B cells. *Current Opinion in Immunology* **16**, 304–313.

SMITH, E. & SIGVARDSSON, M. (2004). The roles of transcription factors in B lymphocyte commitment, development, and transformation. *Journal of Leukocyte Biology* **75**, 973–981.

SZABO, S. J., KIM, S. T., COSTA, G. L., ZHANG, X., FATHMAN, C. G. & GLIMCHER, L. H. (2000). A novel transcription factor, T-bet, directs Th1 lineage commitment. *Cell* **100**, 655–669.

TAKEDA, K. & AKIRA, S. (2004). TLR signaling pathways. *Seminars in Immunology* **16**, 3–9.

TAKEDA, K., KAISHO, T. & AKIRA, S. (2003). Toll-like receptors. *Annual Reviews in Immunology* **21**, 335–376.

TOSHCHAKOV, V., JONES, B. W., PERERA, P. Y., THOMAS, K., CODY, M. J., ZHANG, S., WILLIAMS, B. R., MAJOR, J., HAMILTON, T. A., FENTON, M. J. & VOGEL, S. N. (2002). TLR4, but not TLR2, mediates IFN-beta-induced STAT1alpha/beta-dependent gene expression in macrophages. *Nature Immunology* **3**, 392–398.

WAKELAM, M. J. & HARNETT, M. M. (1998). Phospholipase A2 (EC 3.1.1.4) and D (EC 3.1.4.4) signalling in lymphocytes. *Proceedings of the Nutrition Society* **57**, 551–554.

ZHENG, W. & FLAVELL, R. A. (1997). The transcription factor GATA-3 is necessary and sufficient for Th2 cytokine gene expression in CD4 T cells. *Cell* **89**, 587–596.

ZHOU, M. & OUYANG, W. (2003). The function role of GATA-3 in Th1 and Th2 differentiation. *Immunology Research* **28**, 25–37.

Evasion of innate immunity by vaccinia virus

I. R. HAGA* *and* A. G. BOWIE

Department of Biochemistry, Trinity College, Dublin 2, Ireland

SUMMARY

Vaccinia virus, a member of the Poxviridae, expresses many proteins involved in immune evasion. In this review, we present a brief characterisation of the virus and its effects on host cells and discuss representative secreted and intracellular proteins expressed by vaccinia virus that are involved in modulation of innate immunity. These proteins target different aspects of the innate response by binding cytokines and interferons, inhibiting cytokine synthesis, opposing apoptosis or interfering with different signalling pathways, including those triggered by interferons and toll-like receptors.

Key words: vaccinia virus, immune evasion, signal transduction, nuclear factor kappa B, toll-like receptors.

INTRODUCTION

Vaccinia virus (VV) was the live vaccine used to achieve the global eradication of smallpox, a devastating human disease caused by variola virus (Fenner, 1988). VV is a member of the *Poxviridae*. Poxviruses are characterized by having a large ovoid or brick-shaped virion containing enzymes and factors for messenger RNA (mRNA) synthesis and a single linear double stranded DNA (dsDNA) genome of 130–300 kilobase pairs (kb) (Moss, 2001). Replication of poxviruses occurs entirely in the cell cytoplasm. The *Poxviridae* is divided in sub-families *Entomopoxvirinae* (insect poxviruses) and *Chordopoxvirinae* (vertebrate poxviruses). Generally, members of the same genus have similar virion morphology and present extensive serological cross-reaction and cross protection in laboratory animals (Fenner, 1988).

VV is the prototypic member of the *Orthopoxvirus* genus (sub-family *Chordopoxvirinae*) and remains its most extensively studied member. It was the first mammalian virus to be visualised microscopically, grown in tissue culture, titrated accurately, purified physically and analysed biochemically (Moss, 2001). VV can infect a broad range of mammalian species and can grow in many cell lines *in vitro* (Fenner, Wittek & Dumbell, 1989). The highly attenuated VV strain modified vaccinia Ankara (MVA) was derived from strain Ankara by more than 570 passages in chicken embryo fibroblasts (Hochstein-Mintzel, Huber & Stickl, 1972; Mayr *et al.* 1978) and was used as a smallpox vaccine in over 120 000 individuals without complications (Stickl *et al.* 1974; Mayr *et al.* 1978; Mahnel & Mayr, 1994).

In this review, we present a brief description of VV biology and the host innate response to viruses, before focusing on some of the VV genes involved in evading this host response, with particular reference to recently described VV proteins that modulate toll-like receptor (TLR)-mediated signalling.

OVERVIEW OF VV LIFE CYCLE

VV produces two forms of infectious progeny, intracellular mature virus (IMV) and extracellular enveloped virus (EEV). IMV represents the majority of infectious progeny and remains inside the infected cell until cell lysis (Appleyard, Hapel & Boulter, 1971; Ichihashi, Matsumoto & Dales, 1971). Although EEV is only a small proportion of the progeny virus, it is important biologically. It is released into the extracellular space before cell death (Appleyard *et al.* 1971). The EEV contains an additional lipid membrane, which contains cellular proteins and at least 5 VV-encoded polypeptides that are absent from IMV (Smith & Vanderplasschen, 1998; Vanderplasschen, Hollinshead & Smith, 1998; Smith, Vanderplasschen & Law, 2002).

Due to their large size (approximately 350×270 nm) VV virions are discernible by light microscopy in infected cells and were the first viruses to be observed microscopically. In traditional electron microscopy analyses of infected cells, the virions appear brick-shaped and consist of an internal electron-dense, biconcave core (containing the virus genome, major structural proteins and virion enzymes and two lateral bodies on each of its sides) (Moss, 2001).

The surface of the core contains closely packed, regularly arranged spikes, which comprise the palisade (Dubochet *et al.* 1994). The core structure is wrapped by a lipid membrane and additional virus proteins, which collectively constitute the IMV

* Corresponding author: Ismar R. Haga, Department of Biochemistry, Trinity College, Dublin 2, Ireland. Tel: +353 1 608 3047, fax: +353 1 677 2400, E-mail: ismar. haga@tcd.ie

Parasitology (2005), **130**, S11–S25. © 2005 Cambridge University Press
doi:10.1017/S0031182005008127 Printed in the United Kingdom

membrane. Morphologically, EEV particles are similar to IMV, but contain an additional lipid membrane, which is extremely loose and fragile (Stokes, 1976; Roos *et al.* 1996). The presence of this extra lipid membrane and associated proteins makes EEV structurally and biologically distinct from IMV. EEV mediates the dissemination of virus within the infected host and protection is better achieved by immunity to EEV proteins than IMV, although immunity to IMV can still be helpful (Boulter, 1969; Appleyard *et al.* 1971; Boulter *et al.* 1971; Boulter & Appleyard, 1973; Payne, 1980; Law, Putz & Smith, 2005).

Like all the other members of the *Orthopoxvirus* genus, the VV genome is a dsDNA molecule of approximately 190 kb (length varies according to virus strain). Generally, genes from the central region of the genome are well conserved, are mostly essential for virus replication and are transcribed from either DNA strand. In contrast, genes located near the termini are more diverse, non-essential for virus replication and are generally transcribed towards the termini.

EFFECTS OF POXVIRUS INFECTION ON THE HOST CELL

Infection with VV results in profound changes in cell function, morphology and metabolism, which are termed cytopathic effect (CPE). Visible effects *in vitro* include cell rounding and detachment from neighbouring cells (Bablanian, 1970; Bablanian *et al.* 1978), alterations to the actin cytoskeleton (Hiller *et al.* 1979; Cudmore *et al.* 1995), microtubules (Ploubidou *et al.* 2000) and membrane permeability (Carrasco & Esteban, 1982). This is followed by a phase of enhanced cell migration involving the sequential elongation and condensation of lamellopodia from the cell body (Sanderson, Way & Smith, 1998). Other distinctive changes in cell morphology late during infection include the formation of virus-tipped microvilli (Stokes, 1976; Cudmore *et al.* 1995) as a result of actin tail extension. VV-induced CPE was traditionally viewed as a gradual degeneration of cell function, morphology and viability. Although this ultimately happens, CPE should not be viewed as a simple shut off of cell functions and random cell degeneration. Some aspects of CPE are a consequence of direct action of viral genes and may constitute a manipulation of cellular mechanisms to the advantage of the virus.

Cellular DNA, mRNA and protein synthesis are inhibited following infection with VV (Buller & Palumbo, 1991). Inhibition of host DNA replication may occur by hydrolysis of nascent single stranded DNA, by a viral endonuclease present in the incoming virion (des Gouttes Olgiati, Pogo & Dales, 1976; Dales, 1990). Concurrent with the orderly expression of VV polypeptides, host protein synthesis decreases gradually after VV infection and is completely shut off around 6 h post-infection (p.i.) (Buller & Palumbo, 1991). In the absence of VV early gene expression, IMV surface tubules (Mbuy, Morris & Bubel, 1982), the B1R protein kinase (Beaud *et al.* 1994) and F17R phosphoprotein (Person-Fernandez & Beaud, 1986) have been shown to inhibit host protein translation. Following expression of VV early genes, translation of cellular mRNA is inhibited selectively by small non-translated polyadenylated mRNAs (Bablanian *et al.* 1991; Lu & Bablanian, 1996). Additionally, the overall population of cellular mRNAs is reduced progressively, being replaced by VV mRNAs (Rice & Roberts, 1983), which contributes to the predominant synthesis of VV polypeptides especially at late times p.i. (Moss, 2001). The reduction in the population of host mRNAs, could be due to a decrease in RNA polymerase II activity (and thus cellular mRNA synthesis) and to higher degradation rate of all mRNAs that occurs during VV infection (Rice & Roberts, 1983; Pedley & Cooper, 1984; Buller & Palumbo, 1991).

Although VV has a wide host range, replicating in many cell types of different mammalian and avian species, there are VV host range mutants that fail to replicate in a few specific cell lines (Buller & Palumbo, 1991). VV causes CPE in mouse peritoneal cells or in human leukocytes *in vitro*, but is unable to produce virus progeny from these cells (Nishmi & Bernkopf, 1958). Productive replication of VV, however, occurs if human leukocytes are pre-treated with a mitogen, suggesting that VV requires replicating target cells for efficient replication (Miller & Enders, 1968). More recently it was shown that later stages of VV infection are blocked in primary human macrophages and murine dendritic cells (Broder *et al.* 1994; Bronte *et al.* 1997). Unlike cowpox (CPV), VV is unable to replicate in Chinese hamster ovary (CHO) cells, due to a block in the translation of intermediate mRNAs (Perkus *et al.* 1990). This restriction is caused by a disruption in VV of a host range gene that is present in CPV, encoding the CP77 protein (Spehner *et al.* 1988). In VV WR, two genes, K1L and C7L encode proteins that enable VV replication in human cell lines. The K1L gene product is also necessary for replication in RK-13 cells (rabbit kidney epithelioid cells) (Perkus *et al.* 1990). In a VV K1L deletion mutant, the translation of early VV mRNAs in human or RK-13 cell lines is blocked, and thus there is no VV DNA replication or further gene expression (Ramsey-Ewing & Moss, 1995). Interestingly, the insertion of CP77 into a VV WR K1L deletion mutant partially replaces K1L function, allowing VV to replicate in RK-13 cells (Ramsey-Ewing & Moss, 1996). Therefore, despite the absence of amino acid similarity, CP77, K1L and C7L all influence host range in one cell type or another.

HOST ANTI-VIRAL INNATE IMMUNE RESPONSE

The host immune response to viral infection is biphasic, with innate effectors such as interferons (IFNs), natural killer (NK) cells and macrophages being critical in the early phase, and with adaptive antigen-specific T and B cell responses, which are often essential for clearance of the pathogen and establishment of immunity, developing later. Here, we focus on the innate immune response to VV infection, firstly highlighting its main aspects and then concentrating on the viral strategies to subvert it.

Inflammation is characterized by pain, redness, heat and swelling at the site of infection. The complement system is a crucial innate response that may destroy pathogens directly by lysis or indirectly by opsonising pathogens for phagocytosis by macrophages and neutrophils (Janeway, 2004). Macrophages have an important function as antigen presenting cells (APC) for the activation of T cells and the initiation of a specific immune response. Mice depleted of macrophages are unable to control VV infections due to impaired virus clearance and antigen presentation (Karupiah *et al.* 1996). NK cells are attracted to the site of infection as part of the inflammatory response and kill virus-infected cells, especially cells with reduced levels of MHC class I on their surface (Maudsley & Pound, 1991; See *et al.* 1997). NK cells have a direct cytotoxic activity against VV-infected cells *in vitro* (Brutkiewicz, Klaus & Welsh, 1992) and their *in vivo* depletion is associated with enhanced VV virulence (Bukowski *et al.* 1983).

IFNs are a group of secreted proteins that induce an antiviral state in infected or uninfected cells (Samuel, 1991; Johnson *et al.* 1994). There are two groups of IFNs: type I IFNs are secreted from leukocytes and fibroblasts and include IFN-α and IFN-β. Type I IFNs offer resistance to virus infection, stimulate the expression of MHC class I molecules on the surface of the cell and inhibit cell proliferation (Biron, 1998). Type II IFN, or IFN-γ is secreted from macrophages, NK cells and T lymphocytes. Type II IFN is important in the activation of immune and inflammatory responses and for cell-mediated immunity (Farrar & Schreiber, 1993; Boehm *et al.* 1997). Both type I and type II IFNs are crucial for the restriction of poxviral infections (Schellekens *et al.* 1981; Werenne *et al.* 1985; Rodriguez, Rodriguez & Esteban, 1991; Huang *et al.* 1993; Muller *et al.* 1994; van den Broek *et al.* 1995; Ramshaw *et al.* 1997; Deonarain *et al.* 2000). Mice deficient in IFN receptors are abnormally susceptible to VV infection (van den Broek *et al.* 1995), showing the importance of these molecules in controlling VV infection, and it is therefore not surprising that VV encodes proteins that counteract their activity. The innate immune response also includes the initial inflammatory response that consists of the infiltration of the site of infection with plasma fluids and large numbers of leukocytes. The recruitment of leukocytes to the site of infection is orchestrated by several cytokines and chemokines secreted by the infected cells or resident macrophages. Amongst the inflammatory cells are monocytes that mature into macrophages and which together with neutrophils, the major cell type of early inflammatory infiltrates, may participate in the phagocytosis of infected cells and therefore contain virus infection (Buller & Palumbo, 1991). Importantly, cells present in the inflammatory infiltrate at the site of infection secrete more antiviral and inflammatory cytokines such as tumour necrosis factor (TNF) and IFNs. The complement system of proteins present in the plasma is able to destroy enveloped virions or infected-cells via the classic (Ab-dependent) or alternative (Ab-independent) pathways, which may also have a role in controlling VV infections given that VV has a strategy to avoid both complement pathways (see Table 1).

Apoptosis, programmed cell death (Kerr, Wyllie & Currie, 1972), is a natural and complex process that occurs in response to a variety of stimuli (Osborne, 1996; Jacobson, Weil & Raff, 1997). Apoptosis can be considered as part of the innate immune response because virus infection can induce apoptosis (Everett & McFadden, 1999). This process can also be targeted by poxviruses, as reviewed in Barry & McFadden (2000).

REPRESENTATIVE VV STRATEGIES THAT EVADE THE INNATE IMMUNE RESPONSE

Table 1 shows representative mechanisms utilised by VV to evade the host immune response, which include both secreted and intracellular proteins. Viral immune evasion strategies often target key aspects of anti-viral immunity and are tuned for maximum benefit to the virus. Therefore by studying such strategies one can predict and probe host anti-viral mechanisms. In this section, we focus on some of the strategies used by VV to evade the host innate immunity. However, it should be noted that VV possesses numerous mechanisms to counteract the host immune responses that will not be discussed here. One example is the product of the H1L gene (Fig. 1) (Najarro, Traktman & Lewis, 2001), shown to bind and dephosphorylate STAT1.

Viral semaphorin

Semaphorins are chemoattractants and chemorepellents that are involved in axonal guidance during development of the nervous system. The possibility that semaphorins, with their ability to establish connections between cells, might also function in the immune system was corroborated by the finding that

Table 1. Representative VV genes involved in the evasion of host defence mechanisms

Gene	Function
B18R	Secreted and cell surface bound, binds and inhibits type I IFNs
B8R	Binds and inhibits IFN-γ
B15R	Binds IL-1β, blocks febrile response
A53R[a]	Binds and inhibits TNF-α
K3R[a]	Secreted and cell surface, binds and inhibits TNF-α
B29R/C23L[b]	Binds and inhibits CC chemokines
C21L	Binds C3b and C4b/ inhibits classical and alternative complement pathways
C12L	Binds IL-18 and inhibits IFN-γ production
A39R[c]	Semaphorin, pro-inflammatory
A41L	Anti-inflammatory, unknown ligand
E3L	Binds dsRNA. Prevents activation of PKR and therefore eIF-2 activation and translation arrest. Prevents 2',5'-oligo A synthetase activation, inhibiting RNA degradation by RNAse. Apoptosis inhibitor
K3L	eIF-2α mimic. Blocks eIF-2α phosphorylation and PKR autophosphorylation, inhibiting translational arrest
B22R	Serpin
B13R	Serpin, inhibits IL-1β converting enzyme (ICE, caspase-I) and therefore production of IL-1β. Also inhibits TNF-α or Fas induced apoptosis
A44L	3β-hydroxysteroid dehydrogenase that synthesizes steroids. Gene deletion leads to attenuation
C2L	Kelch-like protein, contributes to formation of VV-induced cellular projections
F1L	Interferes with release of cytochrome c
K1L	Prevents IκB degradation
N1L	Homodimer, targets IKK complex
A52R	Targets IRAK2 and TRAF6
A46R	Targets multiple TIR adaptors

Genes indicated are from VV WR, except: a (VV Lister, Evans, USSR); b (VV Lister, Evans); c (VV Copenhagen). Clear area: secreted gene products, shaded area: intracellular gene products. Adapted from Smith *et al.* 1997; Alcami & Koszinowski, 2000, with modifications.

poxvirus vSEMA (encoded by VV A39R in the Western Reserve (WR) strain) binds to a virus-encoded semaphorin protein receptor (VESPR) in macrophages and induces the expression of cytokines and adhesion molecules (Comeau *et al.* 1998). Expression of vSEMA by VV had no effect in a mouse model of systemic infection, but had some pro-inflammatory activity in an intradermal model (Gardner *et al.* 2001). Recently, A39R was shown to bind plexin C1 (a semaphorin receptor) in dendritic cells (DCs), inhibiting integrin-mediated adhesion and spreading *in vitro*. This was followed by a decrease in integrin signalling and a rearrangement of the actin cytoskeleton (Walzer, Galibert & De Smedt, 2005).

Binding soluble cytokines and interferons and their receptors

The first viral cytokine-receptor to be identified was a viral soluble TNF receptor (vTNFR) (encoded by A53R, see Table 1) (Smith *et al.* 1991), soon followed by viral soluble receptors for IL-1β (vIL-1βR) (encoded by B15R, Fig. 1) (Alcami & Smith, 1992) and IFN-γ (vIFN-γR) (encoded by B8R) (Upton, Mossman & McFadden, 1992). These viral genes encode proteins with sequence similarity to the extracellular binding domains of host cytokine receptors, but they lack the transmembrane and

signalling domains. They were shown to bind cytokines with high affinity and to neutralise their activity (Smith, 2000).

This viral strategy mimics the expression of host soluble versions of cytokine receptors. These receptors are either cleaved from the cell surface or generated by alternative splicing. Cytokine-binding proteins, which modulate cytokine activity, are also present. However, the viral counterparts have unique properties, possibly because they have been optimised during virus-host evolution to be potent cytokine inhibitors (Alcami, 2003). Some viral proteins have binding specificities that are different from those of the cellular receptors. The mammalian type I IL-1R binds IL-1α, IL-1β and IL-1 receptor antagonist (IL-1ra), but VV B15R is specific for IL-1β (Alcami & Smith, 1992). Further studies on B15R demonstrated that IL-1β, and not other cytokines, is the major endogenous pyrogen in a poxvirus infection (Alcami & Smith, 1996).

Other viral soluble cytokine-binding proteins have limited or no sequence similarity to cellular counterparts (Symons, Alcami & Smith, 1995). A common folding of the immunoglobulin domains in the viral IFN-α/β receptors or binding protein (vIFN-α/β-BP) (encoded by B18R) and of the fibronectin type III domains of the cellular counterpart might account for their interaction with the same ligands. VV vIFN-α/β-BP, like vIFN-γR, binds IFNs

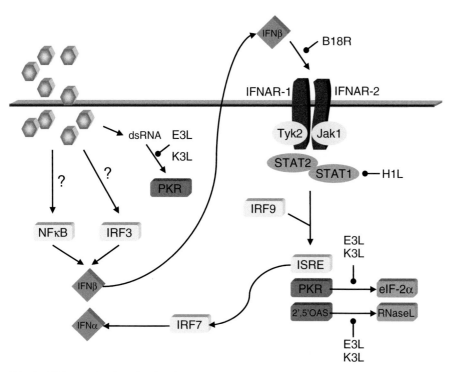

Fig. 1. VV proteins involved in blocking type I IFN and PKR. Please see text for details about E3L, K3L and B18R. STAT1 was shown to be a substrate for H1L in experiments with IFNγ. Although experiments to determine its effect on type I IFN signalling were not carried out, it is possible that H1L might also block this signalling via dephosphorylation of STAT1. Abbreviations: dsRNA, double stranded RNA; PKR, dsRNA-dependent protein kinase; IRF, interferon-regulatory factor; STAT, signal transducer and activator of transcription; ISRE, interferon-stimulated response element; IFN, interferon; IFNAR, interferon α/β receptor; Jak, Janus kinase; Tyk, tyrosine kinase; eIF-2α, eukaryotic translation initiation factor 2α; NFκB, nuclear factor kappa B.

from several species (Symons *et al.* 1995). It can also bind to the cell surface after secretion as a mechanism to prevent IFN from binding to cellular IFN-α/βRs and so protect cells from IFN in solution and on the cell surface (Alcami, Symons & Smith, 2000).

Subsequently, a family of secreted proteins encoded by a number of poxviruses (VV (encoded by C12L, Fig. 2), CPV, molluscum contagiosum virus (MCV), ectromelia virus (ECV) and variola virus) that bind IL-18 (vIL-18BPs) and block IL-18 activity (Xiang & Moss, 1999; Born *et al.* 2000; Smith, Bryant & Alcami, 2000) have been identified. The vIL-18-BPs have no sequence similarity to the mammalian IL-18Rs that mediate the biological effects of IL-18, but the viral proteins are related to human and mouse IL-18BPs that inhibit IL-18 activity (Aizawa *et al.* 1999; Novick *et al.* 1999). The vIL-18BP that is encoded by MCV has an extended carboxy-terminal region that is not found in the IL-18BPs that are encoded by VV, CPV, ECV, humans or mice (Smith *et al.* 2000).

Blocking interleukin synthesis

In addition to the blocking of IL-1β by the IL-1β soluble viral receptor, IL-1β synthesis in VV-infected cells is prevented by one of the poxviral

encoded serine protease inhibitors (serpins) (encoded by B13R and B22R, Fig. 2). Serpins are a group of enzymes characterized by having a highly reactive serine residue in their active site. Serpins regulate, intra- or extracellularly, the function of several proteolytic enzymes involved in diverse physiological processes such as complement activation, inflammation or apoptosis. VV serpin SPI-2 (encoded in VV WR by B13R) and its CPV counterpart cytokine response modifier A (crmA) block caspase-1, a cysteine proteinase also known previously as IL-1β converting enzyme (ICE). Caspase-1 cleaves pro-IL-1β and pro-IL-18 precursors to produce IL-1β and IL-18, respectively (Thornberry *et al.* 1992; Dinarello, 1999). Deletion of B13R in VV did not affect virulence or the febrile response (controlled in part by systemic levels of IL-1β) in mice infected intranasally with VV indicating that: (1) IL-1β produced by infected cells contributes minimally to the systemic levels of IL-1β and (2) the inhibition of febrile response in VV infected mice is primarily due to the activity of the viral IL-1β soluble receptor (Kettle *et al.* 1997). The major role of SPI-2 in VV infection is to block apoptosis (see below).

Recent immunogenicity studies have been performed using sets of viruses lacking B13R or B22R and expressing exogenous proteins. Initially, the

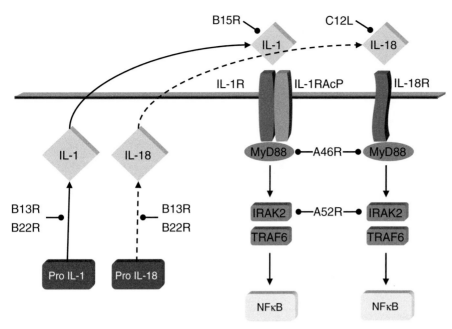

Fig. 2. VV proteins involved in blocking IL-1 and IL-18. Please see text for details about B13R, B22R, B15R, C12L, A46R and A52R. Abbreviations: IL, interleukin; IRAK, IL-1R associated protein kinase; TRAF, tumour necrosis factor receptor-associated factor; IL-1R, IL-1 receptor; IL-1RacP, IL-1 receptor associated protein; IL-18R, IL-18 receptor; NFκB, nuclear factor kappa B; MyD88, myeloid differentiation factor 88.

deletion mutant viruses were engineered to co-express the glycoprotein of vesicular stomatitis virus (VSV) (Legrand *et al.* 2004). In immunocompromised mice, deletion of either gene decreases viral replication and significantly extends time of survival. In immunocompetent mice, deletion of either gene attenuates the virus and elicits potent humoral, T-helper and cytotoxic T-cell immune responses. Deletion mutant viruses lacking B13R or B22R and co-expressing IFN-γ replicated to high titres in tissue culture but were avirulent in both immunocompetent and immunocompromised mice (Legrand *et al.* 2005). The authors suggest that IFN-γ co-expression and the inactivation of one or more VV immune-modulating genes provide an optimised method for increasing the safety while maintaining efficacy of VV vaccines.

Opposing apoptosis

Apoptosis acts as an innate immune response against many invading pathogens including viruses. The profound intracellular alterations that accompany apoptosis, such as the shut down of the macro-molecular synthetic machinery crucial for virus replication, can be effective anti-viral mechanisms particularly if apoptosis is triggered early during infection. In this case, apoptosis could prevent completion of virus replication. To counteract this, several virus families, including the poxviruses, encode anti-apoptotic proteins (Barry & McFadden, 2000). One such protein is CP77 from CPV, already mentioned as a host-range protein. CHO cells

infected by VV in the absence of CP77 undergo all the morphological changes characteristic of apoptosis, but these are delayed when CP77 is expressed, allowing progression of the VV replication cycle (Ramsey-Ewing & Moss, 1998). Another example of a poxviral anti-apoptotic protein is the E3L gene product that prevents the activation of dsRNA-dependent PKR and consequent translational arrest, induced by IFN (see below). Deletion of E3L leads to a narrowing of host range and an accelerated apoptosis in restricted host cells, although it is not known whether this is a consequence of PKR inhibition by E3L or the effect of E3L on other cellular proteins critical for apoptosis induction (Kibler *et al.* 1997).

As mentioned above, the VV WR SPI-2 is important in inhibiting apoptosis triggered by TNF-α or Fas ligand (FasL). Binding of soluble TNF or FasL (present for instance on the cell surface of CTLs or NK cells) to cellular TNF receptor 1 (TNFR1) or Fas, respectively, triggers the activation of the caspase cascade leading to proteolytic cleavage of target proteins. These ultimately produce the phenotype of apoptotic cells, including nuclear fragmentation, loss of cell adhesion (in adherent cells) and cell shrinkage (Zimmermann, Bonzon & Green, 2001). Apoptotic cells are phagocytosed prior to the release of intracellular contents and without an inflammatory response. Although SPI-2/crmA is a potent inhibitor of caspase-1, the prototypic enzyme of the caspase family, it is likely that the SPI-2 blocking of apoptosis is due to the inhibition of another caspase, because caspase-1 (and caspase-1 related caspases) are predominantly involved with inflammation and cytokine activation rather than pro-apoptotic activity (Barry & McFadden, 2000). A possible target is caspase-8, which is involved in progression to apoptosis (Zhou *et al.* 1997). The inhibition of TNF-induced apoptosis by intracellular SPI-2 is another mechanism by which VV blocks TNF activity in addition to the expression of soluble TNF receptors.

Another serpin (SPI-1) encoded by VV WR B22R has counterparts in rabbitpox and CPV. Deletion of SPI-1 in rabbitpox virus inhibited virus replication in human lung carcinoma cell lines (A549) due to premature apoptosis (Ali *et al.* 1994; Brooks *et al.* 1995). Transient expression of the WR SPI-1 in HeLa cells, however, gave no protection against Fas or TNFR-mediated apoptosis (Dobbelstein & Shenk, 1996), neither did rabbitpox or CPV SPI-1 protein protect against Fas-mediated CTL killing (Macen *et al.* 1996). Complete inhibition of Fas-mediated apoptosis and CTL granule killing however did require both rabbitpox SPI-1 and SPI-2, indicating that SPI-1 has a role in blocking apoptosis and can act in synergy with SPI-2 (Macen *et al.* 1996). Like deletion of SPI-2 (ΔB13R), deletion of SPI-1 (ΔB22R) from VV WR had no effect in virus replication *in vitro* or virulence in an intranasal mouse model of infection (Kettle *et al.* 1995). However, further studies revealed a role for B13R in VV virulence that had not been seen before. In an intranasal model of infection, ΔB13R causes a larger lesion when compared to control wild type and revertant viruses (Tscharke, Reading & Smith, 2002). The enhanced pathology observed in animals infected with ΔB13R was a surprise finding and remains to be reconciled with the *in vitro* studies.

Interfering with PKR

The VV E3L and K3L gene products antagonize the effects of IFNs intracellularly (Davies *et al.* 1992, 1993) (see Fig.1). Binding of IFNs to cellular membrane receptors induces the expression of several enzymes, including the IFN-induced dsRNA-dependent protein kinase (PKR, a pattern recognition receptor (PRR)) and 2′, 5′-oligoadenylate (oligoA) synthetase, which are then activated by dsRNA produced during virus infection. Activated (i.e. autophosphorylated) PKR phosphorylates and thereby inactivates the eukaryotic translation initiation factor 2α (eIF-2α), blocking initiation of protein synthesis. The activated 2′, 5′-oligoA synthetase catalyses the formation of oligoA molecules of various length which in turn activate a latent RNAse (RNAse L, an IFN-induced gene) that prevents protein synthesis by cleavage of RNAs. By binding dsRNA, E3L inhibits the activation of both IFN-induced enzymes, arresting the translational block. The K3L gene product is an eIF-2α mimic that

prevents eIF-2α phosphorylation and also inhibits autophosphorylation of PKR. In conjunction with the secreted viral type I and II IFN receptors, the E3L and K3L gene products contribute to the VV resistance to IFNs. Deletion of either E3L or K3L renders VV more sensitive to IFN *in vitro* and deletion of E3L alone results in a narrowing of host range for VV (Beattie *et al.* 1995*a*; Beattie, Paoletti & Tartaglia, 1995*b*) due to failure to inhibit apoptosis (see above).

The C-terminus of E3L binds to dsRNA, whereas the N-terminus is required for additional regulation of eIF2α phosphorylation. The fact that the N-terminus of E3L is required for full viral pathogenesis in mice (despite the absence of induction of protective immune response (Brandt *et al.* 2005)) suggests an alternative role of eIF2α phosphorylation in regulating viral replication (Langland & Jacobs, 2004).

The crystal structure of K3L has been solved (Dar & Sicheri, 2002), with two regions described as important for PKR binding and inhibitory function. The first, a PKR recognition motif, promotes a high-affinity interaction; the second is a helix insert region. Furthermore, K3L has been recently characterized as one of the VV CD8[+] T cell determinants in C57BL/6 mice (Tscharke *et al.* 2005).

E3L and K3L are also intrinsically linked to VV host range. E3L is required for replication in HeLa cells, but is not required for replication in BHK cells. K3L, on the other hand, is required for replication in BHK cells but not in HeLa cells. Langland & Jacobs (2002) verified that the cell lines varied in the expression of endogenous PKR and that replication of VV in these cell lines led to altered levels of dsRNA synthesis from the virus. VV was able to overcome these cellular variations by regulating PKR activity through the synthesis of either E3L or K3L.

In HeLa cells, E3L is likely to bind and mask virtually all the dsRNA synthesized by the virus at late times post-infection. This sequestering prevents recognition of dsRNA by either PKR or OAS, thereby inhibiting the activation of these enzymes. Deletion of the K3L gene had little effect on replication in HeLa cells. The presence of K3L was unable to rescue replication of virus deleted of the E3L gene. HeLa cells endogenously synthesized relatively large amounts of PKR even in the absence of IFN treatment. In the absence of E3L, PKR could bind to the dsRNA and due to the excess levels of PKR, the K3L product was unable to fully inhibit PKR activation and subsequently led to eIF2α phosphorylation.

The BHK cell line differs from the HeLa cell line in two important ways: firstly, when immunoreactive PKR could be detected in BHK cells, no activation of PKR could be detected under *in vitro* phosphorylation conditions, due to the presence of a cellular PKR inhibitor (Kibler *et al.* 1997); secondly, VV replication in BHK cells seemed to lead to higher levels of dsRNA synthesis than replication in HeLa cells. The authors hypothesized that VV replication in BHK cells led to high enough concentrations of dsRNA such that the E3L gene product could no longer fully sequester all the dsRNA present. Different cell lines differ with regard to endogenous levels of PKR and in the amount of dsRNA made during VV infection. VV might have evolved to encode both E3L and K3L as a means of expanding the viral host range. Depending on the cell line, the necessity of either the E3L or the K3L gene may vary.

Deletion of E3L in MVA results in defective late protein synthesis, viral late transcription and viral DNA replication in HeLa cells (Ludwig *et al.* 2005). Viral early and continuing intermediate transcription associates with degradation of rRNA, indicating rapid activation of 2–5-oligoadenylate synthetase/RNase L in the absence of E3L. The lack of the E3L gene inhibits MVA antigen production in HeLa cells at the level of viral late gene expression and suggests that E3L can prevent activation of additional host factors possibly affecting the MVA molecular life cycle.

PKR is by no means the only PRR involved in detecting viral infections. In recent years, the discovery of toll-like receptors (TLRs) has transformed the fields of innate and adaptive immunity (Iwasaki & Medzhitov, 2004; Pasare & Medzhitov, 2004). The discovery of VV mechanisms of evading TLR-mediated signalling suggests a role for TLRs in recognising VV.

VV STRATEGIES TO BLOCK TLR-MEDIATED SIGNALLING

TLRs are members of the IL-1R/TLR superfamily, which is defined by the presence of a Toll/Interleukin-1 receptor/Resistance (TIR) domain (responsible for mediating downstream signalling). TLRs are PRRs that recognise and respond to specific pathogen-associated molecules (PAMs), such as bacterial lipoproteins and peptidoglycan (TLR2), dsRNA (TLR3), LPS (TLR4), flagellin (TLR5), ssRNA (TLR7 and TLR8) and CpG DNA (TLR9) (reviewed in Takeda, Kaisho & Akira, 2003), and drive the innate and adaptive immune responses, depending on stimuli and cell type. TLRs are required for cellular responses to some viruses and trigger antiviral signalling pathways leading to IFN responses (Rassa & Ross, 2003; Vaidya & Cheng, 2003; Bowie & Haga, 2005).

In mammals, the host defence against microorganisms mainly relies on pathways that originate from the common TIR domain of the TLRs. The TLR family signalling pathway is highly similar to that of the IL-1R family. Both IL-1R and TLRs (with the exception of TLR3) utilise the adaptor

protein myeloid differentiation factor-88 (MyD88) for signalling. MyD88 has a TIR domain in its C-terminal portion but a death domain in its N-terminal portion. Apart from MyD88, there are at least other four mammalian TIR domain-containing adaptors: MyD88-adaptor like (Mal), TIR domain-containing adaptor inducing IFN (TRIF), TRIF-related adaptor molecule (TRAM) and sterile and HEAT–Armadillo motifs (SARM) (reviewed in O'Neill, Fitzgerald & Bowie, 2003).

MyD88 associates with the receptors via homotypic TIR domain interactions. Upon stimulation, MyD88 recruits a death domain-containing serine/threonine kinase, the IL-1R-associated kinase (IRAK) (Martin *et al.* 1994). There are four IRAKs in the human and murine genomes: two with demonstrated kinase activity, IRAK1 and IRAK4, and two inactive kinases, IRAK2 and IRAKM (Janssens & Beyaert, 2003). The exact physiological role of IRAK2 has not been clarified. IRAK1 and IRAK4 are activated by phosphorylation, leading to the dissociation of IRAK1 from the receptor complex. IRAK1 in turn interacts with the tumour necrosis factor receptor-associated factor (TRAF) family member TRAF6, eventually leading to activation of the transcription factor nuclear factor kappa B (NFκB). TRAF6 was initially identified as a signal transducer for IL-1 (Cao *et al.* 1996).

NFκB is a transcription factor that has crucial roles in inflammation, immunity, cell proliferation and apoptosis. The activity of NFκB is regulated by association with IκB, which sequesters NFκB in the cytoplasm until phosphorylated on serine residues by the IκB kinase (IKK) complex (Viatour *et al.* 2005). This phosphorylation targets IκB for degradation, allowing nuclear translocation of NFκB. The IKK complex contains two catalytic subunits, IKKα and IKKβ, as well as a scaffolding protein, IKKγ. A mammalian protein complex that activates IKK was purified and analysed and found to be composed of two subunits: TAK1 and a ubiquitin conjugating enzyme complex (Deng *et al.* 2000). TAK1 is consequently activated via its association with the ubiquitinated TRAF6. Once activated, TAK1 mediates phosphorylation of the IKK complex (Wang *et al.* 2001).

NFκB is not the only transcription factor activated by TLRs. Members of the interferon regulatory factor (IRF) family of transcription factors (Taniguchi *et al.* 2001), namely IRF3, IRF5 and IRF7, are also activated by TLRs. Specifically, TLR3 and TLR4, via TRIF, activate IRF3, while TLR7 and TLR8, via MyD88, activate IRF5 and IRF7 (Wietek *et al.* 2003; Takaoka *et al.* 2005; Uematsu *et al.* 2005). IRFs are transcriptional activators for IFNα, IFNβ and IFN-stimulated genes and can act as direct transducers of virus-mediated signaling pathways activating IFNα and IFNβ in infected cells (Taniguchi *et al.* 2001).

We have identified two VV proteins that can antagonise IL-1R and TLR signalling (Bowie *et al.* 2000). A database search for VV-encoded proteins containing putative TIR domains initially identified A46R. A52R was then identified as the highest scoring sequence with significant similarity (A46R and A52R share 16·3% amino acid identity and 32·9% amino acid similarity). Neither A46R nor A52R have signal sequences, or potential N- and O-glycosylation sites.

Preliminary studies showed that overexpression of both proteins could block IL-1-mediated NFκB activation (representing a third VV mechanism of counteracting IL-1, Fig. 1), and A52R could also block NFκB activation triggered by TLR4 (Bowie *et al.* 2000). Further experiments revealed that A52R actually blocks NFκB activation by multiple TLRs and associates with both IRAK2 and TRAF6, two molecules directly involved in TLR signalling. These associations disrupt signalling complexes containing these proteins (Harte *et al.* 2003). A mutant virus in which the A52R gene was deleted was attenuated *in vivo* when compared to control wild type and revertant viruses. The role of A52R in VV virulence is at least partially explained by the inhibition of multiple pathways leading to NFκB activation. Interestingly, a peptide derived from A52R was shown to inhibit *in vitro* TLR-induced cytokine secretion and reduced bacterial-induced inflammation in a mouse model of otitis media with effusion (McCoy *et al.* 2005).

Recently, we have demonstrated that A46R can target multiple TLR adaptors (namely MyD88, Mal, TRIF and TRAM) and interfere with downstream activation of MAP kinases and NFκB. A46R inhibited TRIF-dependent signalling, including IRF3 activation and TRIF-dependent gene induction (Stack *et al.* 2005). Considering the potential role of TRIF in controlling VV replication in macrophages (VV replicated to a higher titre in peritoneal macrophages from TRIF –/– knockout mice when compared to heterozygous and wild type mice, suggesting a role for TRIF in inhibiting VV replication) (Hoebe *et al.* 2003) and the attenuation caused by the deletion of A46R from the VV genome, the targeting of this molecule by A46R may represent a novel and particularly important viral immune evasion mechanism.

Furthermore, A46R and A52R were shown to be non-redundant. A52R potently inhibited poly(I:C)/TLR3-induced NFκB activation, whereas A46R had little effect. This contrasts with the finding that A46R, but not A52R, is a strong inhibitor of poly(I:C)/TLR3-induced IRF3 activation. Also, only A46R was able to inhibit TLR-induced MAPK activation (Stack *et al.* 2005).

Another VV protein recently shown to inhibit NFκB is N1L. Originally identified in concentrated supernatant of VV-infected cell cultures, the product

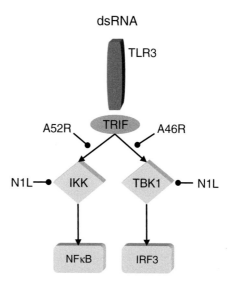

Fig. 3. Proposed model for the inhibition of TLR3/TRIF-mediated signalling by VV. Please see text for details about A46R, A52R and N1L. Abbreviations, TLR, Toll-like receptor; TRIF, TIR domain-containing adaptor-inducing IFN; IKK, IκB kinase; TBK1, TANK-binding kinase 1; NFκB, nuclear factor kappa B; IRF, interferon-regulatory factor.

of the N1L gene is actually an intracellular protein and exists as an homodimer in physiological conditions (Bartlett *et al.* 2002). The N1L gene is not essential for virus replication *in vitro* and its deletion does not affect growth properties in cell culture. Viruses lacking N1L are strongly attenuated *in vivo*, both in the intradermal and in the intranasal models of infection (Bartlett *et al.* 2002). This is in contrast to a variety of other VV immunomodulators, associated with virulence only in one of these two routes (Tscharke *et al.* 2002).

Further studies with the overexpression of N1L in mammalian cells revealed that N1L inhibits NFκB activation by a variety of stimuli, including IL-1β, TNFα and TLR2, TLR3 and TLR4 agonists (DiPerna *et al.* 2004). NFκB activation by different TIR adaptors was also shown to be inhibited by N1L, pointing to a putative N1L target downstream in the signalling pathway. This was confirmed by experimental evidence showing that N1L associates with several proteins of the IKK complex (the point of convergence for all pathways inhibited by N1L), with TBK1 being the most likely target of N1L.

A final identified VV inhibitor of NFκB is K1L. Following the finding that MVA infection induces NFκB activation, subsequent experiments revealed that the insertion of the K1L gene into the MVA genome was sufficient for inhibiting degradation of host IκBα (Shisler & Jin, 2004). Transcription of both the NFκB-regulated cellular TNF gene (regulated by a natural promoter) and a luciferase gene (regulated by an artificial promoter) was repressed by infection of MVA expressing K1L. K1L is necessary for inhibiting NFκB activation in RK-13 cells

(Shisler & Jin, 2004), as well as being a host range factor in this cell line (see above). The authors did not imply a link between host range and NFκB activation nor did they investigate whether MVA-mediated NFκB activation is TLR-dependent. K1L may inhibit IκBα degradation by interacting with IKK to prevent phosphorylation or by hampering kinases that act upstream of IKK. The targets of K1L have not yet been identified.

FINAL REMARKS

For several reasons, the NFκB pathway provides an attractive target for viral pathogens. Activation of NFκB is a rapid, immediate-early event that occurs within minutes after exposure to a relevant inducer, does not require *de novo* protein synthesis, and results in a strong transcriptional stimulation of several early viral genes as well as cellular genes (Hiscott, Kwon & Genin, 2001). Many viruses have evolved mechanisms to target the NFκB pathway to facilitate their replication, cell survival, and evasion of immune responses. In addition, some viruses use the NFκB pathway either for its anti-apoptotic properties to evade the host defence mechanisms or to trigger apoptosis as a mechanism of virus spread (Hiscott *et al.* 2001).

NFκB is known to be activated by multiple viruses, including HIV-1, HTLV-1, hepatitis B virus (HBV), hepatitis C virus (HCV), Epstein-Barr virus (EBV), influenza virus (Hiscott *et al.* 2001) and respiratory syncytial virus (RSV) (Kurt-Jones *et al.* 2000). Poxviruses, however, seem to favour the inhibition of NFκB. At least four different VV-encoded proteins are capable of inhibiting NFκB activation – and three of those (A46R, A52R and N1L) specifically block multiple TLR-induced signals, suggesting that TLRs might have a role in sensing VV. Recently, many TLRs have been linked to sensing different viruses (for reviews, see (Rassa & Ross, 2003; Vaidya & Cheng, 2003; Bowie & Haga, 2005)), and the ligands for the TLRs in question can be either viral proteins or nucleic acids, we are faced with important questions: does any TLR actually sense VV and, if so, which TLR recognise which viral component? Considering the characteristics of poxviruses, we could speculate that the most likely TLRs to be involved would be TLR3 (which senses dsRNA) and TLR9 (which can sense viral dsDNA genomes) (Lund *et al.* 2003). TLR3, TLR7, TLR8 and TLR9 are all intracellular receptors that recognise nucleic acids, with TLR7 and TLR8 linked to ssRNA (Diebold *et al.* 2004; Heil *et al.* 2004; Lund *et al.* 2004). However, other TLRs could not be excluded. For instance, VV also targets TLR4 signalling (Stack *et al.* 2005) and this receptor is known to be activated by proteins from other viruses (Kurt-Jones *et al.* 2000; Burzyn *et al.* 2004; Triantafilou & Triantafilou, 2004). Fig. 3 shows a hypothetical model for the sensing of VV by TLR3 and the known virus-encoded proteins able to block the triggered signalling pathways. Apart from the fact that VV is able to block every single one of the pathways depicted, it should be noted that blocking of NFκB activation by different proteins occurs at different points in the signalling cascade.

In conclusion, VV possesses many strategies to evade the host innate immune response. Recent work with TLRs support the hypothesis that these receptors are involved in sensing and responding to VV infection.

ACKNOWLEDGMENTS

This work was supported by Science Foundation Ireland (SFI).

REFERENCES

AIZAWA, Y., AKITA, K., TANIAI, M., TORIGOE, K., MORI, T., NISHIDA, Y., USHIO, S., NUKADA, Y., TANIMOTO, T., IKEGAMI, H., IKEDA, M. & KURIMOTO, M. (1999). Cloning and expression of interleukin-18 binding protein. *FEBS Letters* **445**, 338–342.

ALCAMI, A. (2003). Viral mimicry of cytokines, chemokines and their receptors. *Nature Reviews Immunology* **3**, 36–50.

ALCAMI, A. & KOSZINOWSKI, U. H. (2000). Viral mechanisms of immune evasion. *Trends in Microbiology* **8**, 410–418.

ALCAMI, A. & SMITH, G. L. (1992). A soluble receptor for interleukin-1 beta encoded by vaccinia virus: a novel mechanism of virus modulation of the host response to infection. *Cell* **71**, 153–167.

ALCAMI, A. & SMITH, G. L. (1996). A mechanism for the inhibition of fever by a virus. *Proceedings of the National Academy of Sciences, USA* **93**, 11029–11034.

ALCAMI, A., SYMONS, J. A. & SMITH, G. L. (2000). The vaccinia virus soluble alpha/beta interferon (IFN) receptor binds to the cell surface and protects cells from the antiviral effects of IFN. *Journal of Virology* **74**, 11230–11239.

ALI, A. N., TURNER, P. C., BROOKS, M. A. & MOYER, R. W. (1994). The SPI-1 gene of rabbitpox virus determines host range and is required for hemorrhagic pock formation. *Virology* **202**, 305–314.

APPLEYARD, G., HAPEL, A. J. & BOULTER, E. A. (1971). An antigenic difference between intracellular and extracellular rabbitpox virus. *Journal of General Virology* **13**, 9–17.

BABLANIAN, R. (1970). Studies on the mechanism of vaccinia virus cytopathic effects: effect of inhibitors of RNA and protein synthesis on early virus-induced cell damage. *Journal of General Virology* **6**, 221–230.

BABLANIAN, R., ESTEBAN, M., BAXT, B. & SONNABEND, J. A. (1978). Studies on the mechanisms of vaccinia virus cytopathic effects. I. Inhibition of protein synthesis in infected cells is associated with virus-induced RNA synthesis. *Journal of General Virology* **39**, 391–402.

BABLANIAN, R., GOSWAMI, S. K., ESTEBAN, M., BANERJEE, A. K. & MERRICK, W. C. (1991). Mechanism of selective translation of vaccinia virus mRNAs: differential role of poly(A) and initiation factors in the translation of viral and cellular mRNAs. *Journal of Virology* **65**, 4449–4460.

BARRY, M. & McFADDEN, G. (2000). Regulation of Apoptosis by Poxviruses. In *Effects of Microbes on the Immune System* (Eds. Cunningham, M. W. & Fujinami, R. S.), Chapter 31, Lippincott Williams & Wilkins, Philadelphia, pp. 509–520.

BARTLETT, N., SYMONS, J. A., TSCHARKE, D. C. & SMITH, G. L. (2002). The vaccinia virus N1L protein is an intracellular homodimer that promotes virulence. *Journal of General Virology* **83**, 1965–1976.

BEATTIE, E., DENZLER, K. L., TARTAGLIA, J., PERKUS, M. E., PAOLETTI, E. & JACOBS, B. L. (1995a). Reversal of the interferon-sensitive phenotype of a vaccinia virus lacking E3L by expression of the reovirus S4 gene. *Journal of Virology* **69**, 499–505.

BEATTIE, E., PAOLETTI, E. & TARTAGLIA, J. (1995b). Distinct patterns of IFN sensitivity observed in cells infected with vaccinia K3L- and E3L-mutant viruses. *Virology* **210**, 254–263.

BEAUD, G., SHARIF, A., TOPA-MASSE, A. & LEADER, D. P. (1994). Ribosomal protein S2/Sa kinase purified from HeLa cells infected with vaccinia virus corresponds to the B1R protein kinase and phosphorylates *in vitro* the viral ssDNA-binding protein. *Journal of General Virology* **75**, 283–293.

BIRON, C. A. (1998). Role of early cytokines, including alpha and beta interferons (IFN-alpha/beta), in innate and adaptive immune responses to viral infections. *Seminars in Immunology* **10**, 383–390.

BOEHM, U., KLAMP, T., GROOT, M. & HOWARD, J. C. (1997). Cellular responses to interferon-gamma. *Annual Review of Immunology* **15**, 749–795.

BORN, T. L., MORRISON, L. A., ESTEBAN, D. J., VANDENBOS, T., THEBEAU, L. G., CHEN, N., SPRIGGS, M. K., SIMS, J. E. & BULLER, R. M. (2000). A poxvirus protein that binds to and inactivates IL-18, and inhibits NK cell response. *Journal of Immunology* **164**, 3246–3254.

BOULTER, E. A. (1969). Protection against poxviruses. *Proceedings of the Royal Society of Medicine* **62**, 295–297.

BOULTER, E. A. & APPLEYARD, G. (1973). Differences between extracellular and intracellular forms of poxvirus and their implications. *Progress in Medical Virology* **16**, 86–108.

BOULTER, E. A., ZWARTOUW, H. T., TITMUSS, D. H. & MABER, H. B. (1971). The nature of the immune state produced by inactivated vaccinia virus in rabbits. *American Journal of Epidemiology* **94**, 612–620.

BOWIE, A. G. & HAGA, I. R. (2005). The role of Toll-like receptors in the host response to viruses. *Molecular Immunology* **42**, 859–867.

BOWIE, A., KISS-TOTH, E., SYMONS, J. A., SMITH, G. L., DOWER, S. K. & O'NEILL, L. A. (2000). A46R and A52R from vaccinia virus are antagonists of host IL-1 and toll-like receptor signaling. *Proceedings of the National Academy of Sciences, USA* **97**, 10162–10167.

BRANDT, T., HECK, M. C., VIJAYSRI, S., JENTARRA, G. M., CAMERON, J. M. & JACOBS, B. L. (2005). The N-terminal domain of the vaccinia virus E3L-protein is required for neurovirulence, but not induction of a protective immune response. *Virology* **333**, 263–270.

BRODER, C. C., KENNEDY, P. E., MICHAELS, F. & BERGER, E. A. (1994). Expression of foreign genes in cultured human primary macrophages using recombinant vaccinia virus vectors. *Gene* **142**, 167–174.

BRONTE, V., CARROLL, M. W., GOLETZ, T. J., WANG, M., OVERWIJK, W. W., MARINCOLA, F., ROSENBERG, S. A., MOSS, B. & RESTIFO, N. P. (1997). Antigen expression by dendritic cells correlates with the therapeutic effectiveness of a model recombinant poxvirus tumor vaccine. *Proceedings of the National Academy of Sciences, USA* **94**, 3183–3188.

BROOKS, M. A., ALI, A. N., TURNER, P. C. & MOYER, R. W. (1995). A rabbitpox virus serpin gene controls host range by inhibiting apoptosis in restrictive cells. *Journal of Virology* **69**, 7688–7698.

BRUTKIEWICZ, R. R., KLAUS, S. J. & WELSH, R. M. (1992). Window of vulnerability of vaccinia virus-infected cells to natural killer (NK) cell-mediated cytolysis correlates with enhanced NK cell triggering and is concomitant with a decrease in H-2 class I antigen expression. *Nature Immunology* **11**, 203–214.

BUKOWSKI, J. F., WODA, B. A., HABU, S., OKUMURA, K. & WELSH, R. M. (1983). Natural killer cell depletion enhances virus synthesis and virus-induced hepatitis *in vivo*. *Journal of Immunology* **131**, 1531–1538.

BULLER, R. M. & PALUMBO, G. J. (1991). Poxvirus pathogenesis. *Microbiological Reviews* **55**, 80–122.

BURZYN, D., RASSA, J. C., KIM, D., NEPOMNASCHY, I., ROSS, S. R. & PIAZZON, I. (2004). Toll-like receptor 4-dependent activation of dendritic cells by a retrovirus. *Journal of Virology* **78**, 576–584.

CAO, Z., XIONG, J., TAKEUCHI, M., KURAMA, T. & GOEDDEL, D. V. (1996). TRAF6 is a signal transducer for interleukin-1. *Nature* **383**, 443–446.

CARRASCO, L. & ESTEBAN, M. (1982). Modification of membrane permeability in vaccinia virus-infected cells. *Virology* **117**, 62–69.

COMEAU, M. R., JOHNSON, R., DUBOSE, R. F., PETERSEN, M., GEARING, P., VANDENBOS, T., PARK, L., FARRAH, T., BULLER, R. M., COHEN, J. I., STROCKBINE, L. D., RAUCH, C. & SPRIGGS, M. K. (1998). A poxvirus-encoded semaphorin induces cytokine production from monocytes and binds to a novel cellular semaphorin receptor, VESPR. *Immunity* **8**, 473–482.

CUDMORE, S., COSSART, P., GRIFFITHS, G. & WAY, M. (1995). Actin-based motility of vaccinia virus. *Nature* **378**, 636–638.

DALES, S. (1990). Reciprocity in the interactions between the poxviruses and their host cells. *Annual Review of Microbiology* **44**, 173–192.

DAR, A. C. & SICHERI, F. (2002). X-ray crystal structure and functional analysis of vaccinia virus K3L reveals molecular determinants for PKR subversion and substrate recognition. *Molecular Cell* **10**, 295–305.

DAVIES, M. V., ELROY-STEIN, O., JAGUS, R., MOSS, B. & KAUFMAN, R. J. (1992). The vaccinia virus K3L gene product potentiates translation by inhibiting double-stranded-RNA-activated protein kinase and phosphorylation of the alpha subunit of eukaryotic initiation factor 2. *Journal of Virology* **66**, 1943–1950.

DAVIES, M. V., CHANG, H. W., JACOBS, B. L. & KAUFMAN, R. J. (1993). The E3L and K3L vaccinia virus gene products stimulate translation through inhibition of the double-stranded RNA-dependent protein kinase by different mechanisms. *Journal of Virology* **67**, 1688–1692.

DENG, L., WANG, C., SPENCER, E., YANG, L., BRAUN, A., YOU, J., SLAUGHTER, C., PICKART, C. & CHEN, Z. J. (2000). Activation of the IkappaB kinase complex by TRAF6 requires a dimeric ubiquitin-conjugating enzyme complex and a unique polyubiquitin chain. *Cell* **103**, 351–361.

DEONARAIN, R., ALCAMI, A., ALEXIOU, M., DALLMAN, M. J., GEWERT, D. R. & PORTER, A. C. (2000). Impaired antiviral response and alpha/beta interferon induction in mice lacking beta interferon. *Journal of Virology* **74**, 3404–3409.

DES GOUTTES OLGIATI, D., POGO, B. G. & DALES, S. (1976). Biogenesis of vaccinia: specific inhibition of rapidly labeled host DNA in vaccinia inoculated cells. *Virology* **71**, 325–335.

DIEBOLD, S. S., KAISHO, T., HEMMI, H., AKIRA, S. & REIS E SOUSA, C. (2004). Innate antiviral responses by means of TLR7-mediated recognition of single-stranded RNA. *Science* **303**, 1529–1531.

DINARELLO, C. A. (1999). Interleukin-18. *Methods* **19**, 121–132.

DIPERNA, G., STACK, J., BOWIE, A. G., BOYD, A., KOTWAL, G., ZHANG, Z., ARVIKAR, S., LATZ, E., FITZGERALD, K. A. & MARSHALL, W. L. (2004). Poxvirus protein N1L targets the I-kappaB kinase complex, inhibits signaling to NF-kappaB by the tumor necrosis factor superfamily of receptors, and inhibits NF-kappaB and IRF3 signaling by toll-like receptors. *Journal of Biological Chemistry* **279**, 36570–36578.

DOBBELSTEIN, M. & SHENK, T. (1996). Protection against apoptosis by the vaccinia virus SPI-2 (B13R) gene product. *Journal of Virology* **70**, 6479–6485.

DUBOCHET, J., ADRIAN, M., RICHTER, K., GARCES, J. & WITTEK, R. (1994). Structure of intracellular mature vaccinia virus observed by cryoelectron microscopy. *Journal of Virology* **68**, 1935–1941.

EVERETT, H. & McFADDEN, G. (1999). Apoptosis: an innate immune response to virus infection. *Trends in Microbiology* **7**, 160–165.

FARRAR, M. A. & SCHREIBER, R. D. (1993). The molecular cell biology of interferon-gamma and its receptor. *Annual Review of Immunology* **11**, 571–611.

FENNER, F. (1988). *Smallpox and its Eradication.* World Health Organization, Geneva.

FENNER, F., WITTEK, R. & DUMBELL, K. R. (1989). *The Orthopoxviruses.* Academic Press, London.

GARDNER, J. D., TSCHARKE, D. C., READING, P. C. & SMITH, G. L. (2001). Vaccinia virus semaphorin A39R is a 50–55. kDa secreted glycoprotein that affects the outcome of infection in a murine intradermal model. *Journal of General Virology* **82**, 2083–2093.

HARTE, M. T., HAGA, I. R., MALONEY, G., GRAY, P., READING, P. C., BARTLETT, N. W., SMITH, G. L., BOWIE, A. & O'NEILL, L. A. (2003). The poxvirus protein A52R targets Toll-like receptor signaling complexes to suppress host defense. *Journal of Experimental Medicine* **197**, 343–351.

HEIL, F., HEMMI, H., HOCHREIN, H., AMPENBERGER, F., KIRSCHNING, C., AKIRA, S., LIPFORD, G., WAGNER, H. & BAUER, S. (2004). Species-specific recognition of single-stranded RNA via toll-like receptor 7 and 8. *Science* **303**, 1526–1529.

HILLER, G., WEBER, K., SCHNEIDER, L., PARAJSZ, C. & JUNGWIRTH, C. (1979). Interaction of assembled progeny pox viruses with the cellular cytoskeleton. *Virology* **98**, 142–153.

HISCOTT, J., KWON, H. & GENIN, P. (2001). Hostile takeovers: viral appropriation of the NF-kappaB pathway. *Journal of Clinical Investigation* **107**, 143–151.

HOCHSTEIN-MINTZEL, V., HUBER, H. C. & STICKL, H. (1972). [Virulence and immunogenicity of a modified vaccinia virus (strain MVA) (author's transl).] *Zeitschrift für Immunitatsforschung, Experimentelle und Klinische Immunologie* **144**, 104–156.

HOEBE, K., DU, X., GEORGEL, P., JANSSEN, E., TABETA, K., KIM, S. O., GOODE, J., LIN, P., MANN, N., MUDD, S., CROZAT, K., SOVATH, S., HAN, J. & BEUTLER, B. (2003). Identification of Lps2 as a key transducer of MyD88-independent TIR signalling. *Nature* **424**, 743–748.

HUANG, S., HENDRIKS, W., ALTHAGE, A., HEMMI, S., BLUETHMANN, H., KAMIJO, R., VILCEK, J., ZINKERNAGEL, R. M. & AGUET, M. (1993). Immune response in mice that lack the interferon-gamma receptor. *Science* **259**, 1742–1745.

ICHIHASHI, Y., MATSUMOTO, S. & DALES, S. (1971). Biogenesis of poxviruses: role of A-type inclusions and host cell membranes in virus dissemination. *Virology* **46**, 507–532.

IWASAKI, A. & MEDZHITOV, R. (2004). Toll-like receptor control of the adaptive immune responses. *Nature Immunology* **5**, 987–995.

JACOBSON, M. D., WEIL, M. & RAFF, M. C. (1997). Programmed cell death in animal development. *Cell* **88**, 347–354.

JANEWAY, C. (2004). *Immunobiology 6: The Immune System in Health and Disease.* 6th. Garland Pub., New York.

JANSSENS, S. & BEYAERT, R. (2003). Functional diversity and regulation of different interleukin-1 receptor-associated kinase (IRAK) family members. *Molecular Cell* **11**, 293–302.

JOHNSON, H. M., BAZER, F. W., SZENTE, B. E. & JARPE, M. A. (1994). How interferons fight disease. *Scientific American* **270**, 68–75.

KARUPIAH, G., BULLER, R. M., VAN ROOIJEN, N., DUARTE, C. J. & CHEN, J. (1996). Different roles for CD4 + and CD8 + T lymphocytes and macrophage subsets in the control of a generalized virus infection. *Journal of Virology* **70**, 8301–8309.

KERR, J. F., WYLLIE, A. H. & CURRIE, A. R. (1972). Apoptosis: a basic biological phenomenon with wide-ranging implications in tissue kinetics. *British Journal of Cancer* **26**, 239–257.

KETTLE, S., ALCAMI, A., KHANNA, A., EHRET, R., JASSOY, C. & SMITH, G. L. (1997). Vaccinia virus serpin B13R (SPI-2) inhibits interleukin-1beta-converting enzyme and protects virus-infected cells from TNF- and Fas-mediated apoptosis, but does not prevent IL-1beta-induced fever. *Journal of General Virology* **78**, 677–685.

KETTLE, S., BLAKE, N. W., LAW, K. M. & SMITH, G. L. (1995). Vaccinia virus serpins B13R (SPI-2) and B22R (SPI-1) encode M(r) 38.5 and 40K, intracellular polypeptides that do not affect virus virulence in a murine intranasal model. *Virology* **206**, 136–147.

KIBLER, K. V., SHORS, T., PERKINS, K. B., ZEMAN, C. C., BANASZAK, M. P., BIESTERFELDT, J., LANGLAND, J. O. & JACOBS, B. L. (1997). Double-stranded RNA is a trigger for apoptosis in vaccinia virus-infected cells. *Journal of Virology* **71**, 1992–2003.

KURT-JONES, E. A., POPOVA, L., KWINN, L., HAYNES, L. M., JONES, L. P., TRIPP, R. A., WALSH, E. E., FREEMAN, M. W., GOLENBOCK, D. T., ANDERSON, L. J. & FINBERG, R. W. (2000). Pattern recognition receptors TLR4 and CD14 mediate response to respiratory syncytial virus. *Nature Immunology* **1**, 398–401.

LANGLAND, J. O. & JACOBS, B. L. (2002). The role of the PKR-inhibitory genes, E3L and K3L, in determining vaccinia virus host range. *Virology* **299**, 133–141.

LANGLAND, J. O. & JACOBS, B. L. (2004). Inhibition of PKR by vaccinia virus: role of the N- and C-terminal domains of E3L. *Virology* **324**, 419–429.

LAW, M., PUTZ, M. M. & SMITH, G. L. (2005). An investigation of the therapeutic value of vaccinia-immune IgG in a mouse pneumonia model. *Journal of General Virology* **86**, 991–1000.

LEGRAND, F. A., VERARDI, P. H., JONES, L. A., CHAN, K. S., PENG, Y. & YILMA, T. D. (2004). Induction of potent humoral and cell-mediated immune responses by attenuated vaccinia virus vectors with deleted serpin genes. *Journal of Virology* **78**, 2770–2779.

LEGRAND, F. A., VERARDI, P. H., CHAN, K. S., PENG, Y., JONES, L. A. & YILMA, T. D. (2005). Vaccinia viruses with a serpin gene deletion and expressing IFN-gamma induce potent immune responses without detectable replication *in vivo*. *Proceedings of the National Academy of Sciences, USA* **102**, 2940–2945.

LU, C. & BABLANIAN, R. (1996). Characterization of small nontranslated polyadenylylated RNAs in vaccinia virus-infected cells. *Proceedings of the National Academy of Sciences, USA* **93**, 2037–2042.

LUDWIG, H., MAGES, J., STAIB, C., LEHMANN, M. H., LANG, R. & SUTTER, G. (2005). Role of viral factor E3L in modified vaccinia virus Ankara infection of human HeLa Cells: regulation of the virus life cycle and identification of differentially expressed host genes. *Journal of Virology* **79**, 2584–2596.

LUND, J., SATO, A., AKIRA, S., MEDZHITOV, R. & IWASAKI, A. (2003). Toll-like receptor 9-mediated recognition of Herpes simplex virus-2 by plasmacytoid dendritic cells. *Journal of Experimental Medicine* **198**, 513–520.

LUND, J. M., ALEXOPOULOU, L., SATO, A., KAROW, M., ADAMS, N. C., GALE, N. W., IWASAKI, A. & FLAVELL, R. A. (2004). Recognition of single-stranded RNA viruses by Toll-like receptor 7. *Proceedings of the National Academy of Sciences, USA* **101**, 5598–5603.

MACEN, J. L., GARNER, R. S., MUSY, P. Y., BROOKS, M. A., TURNER, P. C., MOYER, R. W., McFADDEN, G. & BLEACKLEY, R. C. (1996). Differential inhibition of the Fas- and granule-mediated cytolysis pathways by the orthopoxvirus cytokine response modifier A/SPI-2 and SPI-1 protein. *Proceedings of the National Academy of Sciences, USA* **93**, 9108–9113.

MAHNEL, H. & MAYR, A. (1994). [Experiences with immunization against orthopoxviruses of humans and animals using vaccine strain MVA.] *Berliner und Munchener Tierarztliche Wochenschrift* **107**, 253–256.

MARTIN, M., BOL, G. F., ERIKSSON, A., RESCH, K. & BRIGELIUS-FLOHE, R. (1994). Interleukin-1-induced activation of a protein kinase co-precipitating with the type I interleukin-1 receptor in T cells. *European Journal of Immunology* **24**, 1566–1571.

MAUDSLEY, D. J. & POUND, J. D. (1991). Modulation of MHC antigen expression by viruses and oncogenes. *Immunology Today* **12**, 429–431.

MAYR, A., STICKL, H., MULLER, H. K., DANNER, K. & SINGER, H. (1978). [The smallpox vaccination strain MVA: marker, genetic structure, experience gained with the parenteral vaccination and behavior in organisms with a debilitated defence mechanism (author's transl).] *Zentralblatt für Bakteriologie [B]* **167**, 375–390.

MBUY, G. N., MORRIS, R. E. & BUBEL, H. C. (1982). Inhibition of cellular protein synthesis by vaccinia virus surface tubules. *Virology* **116**, 137–147.

McCOY, S. L., KURTZ, S. E., MACARTHUR, C. J., TRUNE, D. R. & HEFENEIDER, S. H. (2005). Identification of a peptide derived from vaccinia virus A52R protein that inhibits cytokine secretion in response to TLR-dependent signaling and reduces *in vivo* bacterial-induced inflammation. *Journal of Immunology* **174**, 3006–3014.

MILLER, G. & ENDERS, J. F. (1968). Vaccinia virus replication and cytopathic effect in cultures in phytohemagglutinin-treated human peripheral blood leukocytes. *Journal of Virology* **2**, 787–792.

MOSS, B. (2001). Poxviridae: the viruses and their replication. In *Fields Virology* (Eds. Knipe, D. M. & Howley, P. M.), Vol. 2, Lippincott Williams & Wilkins, Philadelphia, pp. 2849–2883.

MULLER, U., STEINHOFF, U., REIS, L. F., HEMMI, S., PAVLOVIC, J., ZINKERNAGEL, R. M. & AGUET, M. (1994). Functional role of type I and type II interferons in antiviral defense. *Science* **264**, 1918–1921.

NAJARRO, P., TRAKTMAN, P. & LEWIS, J. A. (2001). Vaccinia virus blocks gamma interferon signal transduction: viral VH1 phosphatase reverses Stat1 activation. *Journal of Virology* **75**, 3185–3196.

NISHMI, M. & BERNKOPF, H. (1958). The toxic effect of vaccinia virus on leucocytes *in vitro*. *Journal of Immunology* **81**, 460–466.

NOVICK, D., KIM, S. H., FANTUZZI, G., REZNIKOV, L. L., DINARELLO, C. A. & RUBINSTEIN, M. (1999). Interleukin-18 binding protein: a novel modulator of the Th1 cytokine response. *Immunity* **10**, 127–136.

O'NEILL, L. A., FITZGERALD, K. A. & BOWIE, A. G. (2003). The Toll-IL-1 receptor adaptor family grows to five members. *Trends in Immunology* **24**, 286–290.

OSBORNE, B. A. (1996). Apoptosis and the maintenance of homoeostasis in the immune system. *Current Opinion in Immunology* **8**, 245–254.

PASARE, C. & MEDZHITOV, R. (2004). Toll-like receptors: linking innate and adaptive immunity. *Microbes and Infection* **6**, 1382–1387.

PAYNE, L. G. (1980). Significance of extracellular enveloped virus in the *in vitro* and *in vivo* dissemination of vaccinia. *Journal of General Virology* **50**, 89–100.

PEDLEY, S. & COOPER, R. J. (1984). The inhibition of HeLa cell RNA synthesis following infection with vaccinia virus. *Journal of General Virology* **65**, 1687–1697.

PERKUS, M. E., GOEBEL, S. J., DAVIS, S. W., JOHNSON, G. P., LIMBACH, K., NORTON, E. K. & PAOLETTI, E. (1990). Vaccinia virus host range genes. *Virology* **179**, 276–286.

PERSON-FERNANDEZ, A. & BEAUD, G. (1986). Purification and characterization of a protein synthesis inhibitor associated with vaccinia virus. *Journal of Biological Chemistry* **261**, 8283–8289.

PLOUBIDOU, A., MOREAU, V., ASHMAN, K., RECKMANN, I., GONZALEZ, C. & WAY, M. (2000). Vaccinia virus infection disrupts microtubule organization and centrosome function. *EMBO Journal* **19**, 3932–3944.

RAMSEY-EWING, A. & MOSS, B. (1995). Restriction of vaccinia virus replication in CHO cells occurs at the stage of viral intermediate protein synthesis. *Virology* **206**, 984–993.

RAMSEY-EWING, A. L. & MOSS, B. (1996). Complementation of a vaccinia virus host-range K1L gene deletion by the nonhomologous CP77 gene. *Virology* **222**, 75–86.

RAMSEY-EWING, A. & MOSS, B. (1998). Apoptosis induced by a postbinding step of vaccinia virus entry into Chinese hamster ovary cells. *Virology* **242**, 138–149.

RAMSHAW, I. A., RAMSAY, A. J., KARUPIAH, G., ROLPH, M. S., MAHALINGAM, S. & RUBY, J. C. (1997). Cytokines and immunity to viral infections. *Immunological Reviews* **159**, 119–135.

RASSA, J. C. & ROSS, S. R. (2003). Viruses and Toll-like receptors. *Microbes and Infection* **5**, 961–968.

RICE, A. P. & ROBERTS, B. E. (1983). Vaccinia virus induces cellular mRNA degradation. *Journal of Virology* **47**, 529–539.

RODRIGUEZ, J. R., RODRIGUEZ, D. & ESTEBAN, M. (1991). Interferon treatment inhibits early events in vaccinia virus gene expression in infected mice. *Virology* **185**, 929–933.

ROOS, N., CYRKLAFF, M., CUDMORE, S., BLASCO, R., KRIJNSE-LOCKER, J. & GRIFFITHS, G. (1996). A novel immunogold cryoelectron microscopic approach to investigate the structure of the intracellular and extracellular forms of vaccinia virus. *EMBO Journal* **15**, 2343–2355.

SAMUEL, C. E. (1991). Antiviral actions of interferon. Interferon-regulated cellular proteins and their surprisingly selective antiviral activities. *Virology* **183**, 1–11.

SANDERSON, C. M., WAY, M. & SMITH, G. L. (1998). Virus-induced cell motility. *Journal of Virology* **72**, 1235–1243.

SCHELLEKENS, H., DE REUS, A., BOLHUIS, R., FOUNTOULAKIS, M., SCHEIN, C., ECSODI, J., NAGATA, S. & WEISSMANN, C. (1981). Comparative antiviral efficiency of leukocyte and bacterially produced human alpha-interferon in rhesus monkeys. *Nature* **292**, 775–776.

SEE, D. M., KHEMKA, P., SAHL, L., BUI, T. & TILLES, J. G. (1997). The role of natural killer cells in viral infections. *Scandinavian Journal of Immunology* **46**, 217–224.

SHISLER, J. L. & JIN, X. L. (2004). The vaccinia virus K1L gene product inhibits host NF-kappaB activation by preventing IkappaBalpha degradation. *Journal of Virology* **78**, 3553–3560.

SMITH, G. L. (2000). Secreted poxvirus proteins that interact with the immune system. In *Effects of Microbes on the Immune System* (Eds. Cunningham, M. W. & Fujinami, R. S.), Lippincott Williams & Wilkins, Philadelphia, pp. 491–507.

SMITH, C. A., DAVIS, T., WIGNALL, J. M., DIN, W. S., FARRAH, T., UPTON, C., McFADDEN, G. & GOODWIN, R. G. (1991). T2 open reading frame from the Shope fibroma virus encodes a soluble form of the TNF receptor. *Biochemical and Biophysical Research Communications* **176**, 335–342.

SMITH, G. L., SYMONS, J. A., KHANNA, A., VANDERPLASSCHEN, A. & ALCAMI, A. (1997). Vaccinia virus immune evasion. *Immunological Reviews* **159**, 137–154.

SMITH, G. L. & VANDERPLASSCHEN, A. (1998). Extracellular enveloped vaccinia virus. Entry, egress, and evasion. *Advances in Experimental Medicine and Biology* **440**, 395–414.

SMITH, G. L., VANDERPLASSCHEN, A. & LAW, M. (2002). The formation and function of extracellular enveloped vaccinia virus. *Journal of General Virology* **83**, 2915–2931.

SMITH, V. P., BRYANT, N. A. & ALCAMI, A. (2000). Ectromelia, vaccinia and cowpox viruses encode secreted interleukin-18-binding proteins. *Journal of General Virology* **81**, 1223–1230.

SPEHNER, D., GILLARD, S., DRILLIEN, R. & KIRN, A. (1988). A cowpox virus gene required for multiplication in Chinese hamster ovary cells. *Journal of Virology* **62**, 1297–1304.

STACK, J., HAGA, I. R., SCHRODER, M., BARTLETT, N. W., MALONEY, G., READING, P. C., FITZGERALD, K. A., SMITH, G. L. & BOWIE, A. G. (2005). Vaccinia virus protein A46R targets multiple Toll-like-interleukin-1 receptor adaptors and contributes to virulence. *Journal of Experimental Medicine* **201**, 1007–1018.

STICKL, H., HOCHSTEIN-MINTZEL, V., MAYR, A., HUBER, H. C., SCHAFER, H. & HOLZNER, A. (1974). [MVA vaccination against smallpox: clinical tests with an attenuated live vaccinia virus strain (MVA) (author's transl).] *Deutsche Medizinische Wochenschrift* **99**, 2386–2392.

STOKES, G. V. (1976). High-voltage electron microscope study of the release of vaccinia virus from whole cells. *Journal of Virology* **18**, 636–643.

SYMONS, J. A., ALCAMI, A. & SMITH, G. L. (1995). Vaccinia virus encodes a soluble type I interferon receptor of novel structure and broad species specificity. *Cell* **81**, 551–560.

TAKAOKA, A., YANAI, H., KONDO, S., DUNCAN, G., NEGISHI, H., MIZUTANI, T., KANO, S., HONDA, K., OHBA, Y., MAK, T. W. & TANIGUCHI, T. (2005). Integral role of IRF-5 in the gene induction programme activated by Toll-like receptors. *Nature* **434**, 243–249.

TAKEDA, K., KAISHO, T. & AKIRA, S. (2003). Toll-like receptors. *Annual Review of Immunology* **21**, 335–376.

TANIGUCHI, T., OGASAWARA, K., TAKAOKA, A. & TANAKA, N. (2001). IRF family of transcription factors as regulators of host defense. *Annual Review of Immunology* **19**, 623–655.

THORNBERRY, N. A., BULL, H. G., CALAYCAY, J. R., CHAPMAN, K. T., HOWARD, A. D., KOSTURA, M. J., MILLER, D. K., MOLINEAUX, S. M., WEIDNER, J. R., AUNINS, J. et al. (1992). A novel heterodimeric cysteine protease is required for interleukin-1 beta processing in monocytes. *Nature* **356**, 768–774.

TRIANTAFILOU, K. & TRIANTAFILOU, M. (2004). Coxsackievirus B4-induced cytokine production in pancreatic cells is mediated through toll-like receptor 4. *Journal of Virology* **78**, 11313–11320.

TSCHARKE, D. C., KARUPIAH, G., ZHOU, J., PALMORE, T., IRVINE, K. R., HAERYFAR, S. M., WILLIAMS, S., SIDNEY, J., SETTE, A., BENNINK, J. R. & YEWDELL, J. W. (2005). Identification of poxvirus CD8+ T cell determinants to enable rational design and characterization of smallpox vaccines. *Journal of Experimental Medicine* **201**, 95–104.

TSCHARKE, D. C., READING, P. C. & SMITH, G. L. (2002). Dermal infection with vaccinia virus reveals roles for virus proteins not seen using other inoculation routes. *Journal of General Virology* **83**, 1977–1986.

UEMATSU, S., SATO, S., YAMAMOTO, M., HIROTANI, T., KATO, H., TAKESHITA, F., MATSUDA, M., COBAN, C., ISHII, K. J., KAWAI, T., TAKEUCHI, O. & AKIRA, S. (2005). Interleukin-1 receptor-associated kinase-1 plays an essential role for Toll-like receptor (TLR)7- and TLR9-mediated interferon-{alpha} induction. *Journal of Experimental Medicine* **201**, 915–923.

UPTON, C., MOSSMAN, K. & McFADDEN, G. (1992). Encoding of a homolog of the IFN-gamma receptor by myxoma virus. *Science* **258**, 1369–1372.

VAIDYA, S. A. & CHENG, G. (2003). Toll-like receptors and innate antiviral responses. *Current Opinion in Immunology* **15**, 402–407.

VAN DEN BROEK, M. F., MULLER, U., HUANG, S., AGUET, M. & ZINKERNAGEL, R. M. (1995). Antiviral defense in mice lacking both alpha/beta and gamma interferon receptors. *Journal of Virology* **69**, 4792–4796.

VANDERPLASSCHEN, A., HOLLINSHEAD, M. & SMITH, G. L. (1998). Intracellular and extracellular vaccinia virions enter cells by different mechanisms. *Journal of General Virology* **79**, 877–887.

VIATOUR, P., MERVILLE, M. P., BOURS, V. & CHARIOT, A. (2005). Phosphorylation of NF-kappaB and IkappaB proteins: implications in cancer and inflammation. *Trends in Biochemical Sciences* **30**, 43–52.

WALZER, T., GALIBERT, L. & DE SMEDT, T. (2005). Poxvirus semaphorin A39R inhibits phagocytosis by dendritic cells and neutrophils. *European Journal of Immunology* **35**, 391–398.

WANG, C., DENG, L., HONG, M., AKKARAJU, G. R., INOUE, J. & CHEN, Z. J. (2001). TAK1 is a ubiquitin-dependent kinase of MKK and IKK. *Nature* **412**, 346–351.

WERENNE, J., VANDEN BROECKE, C., SCHWERS, A., GOOSSENS, A., BUGYAKI, L., MAENHOUDT, M. & PASTORET, P. P. (1985). Antiviral effect of bacterially produced human interferon (Hu-IFN alpha 2) against experimental vaccinia infection in calves. *Journal of Interferon Research* **5**, 129–136.

WIETEK, C., MIGGIN, S. M., JEFFERIES, C. A. & O'NEILL, L. A. (2003). Interferon regulatory factor-3-mediated activation of the interferon-sensitive response element by Toll-like receptor (TLR) 4 but not TLR3 requires the p65 subunit of NF-kappa. *Journal of Biological Chemistry* **278**, 50923–50931.

XIANG, Y. & MOSS, B. (1999). IL-18 binding and inhibition of interferon gamma induction by human poxvirus-encoded proteins. *Proceedings of the National Academy of Sciences, USA* **96**, 11537–11542.

ZHOU, Q., SNIPAS, S., ORTH, K., MUZIO, M., DIXIT, V. M. & SALVESEN, G. S. (1997). Target protease specificity of the viral serpin CrmA. Analysis of five caspases. *Journal of Biological Chemistry* **272**, 7797–7800.

ZIMMERMANN, K. C., BONZON, C. & GREEN, D. R. (2001). The machinery of programmed cell death. *Pharmacology and Therapeutics* **92**, 57–70.

Subversion of host cell signalling by the protozoan parasite *Leishmania*

D. J. GREGORY[1] *and* M. OLIVIER[1,2]*

[1] *Centre for the Study of Host Resistance at the Research Institute of the McGill University Health Centre, and Departments of Microbiology and Immunology*
[2] *Experimental Medicine, McGill University, Montréal, Québec, Canada*

SUMMARY

The protozoa *Leishmania* spp. are obligate intracellular parasites that inhabit the macrophages of their host. Since macrophages are specialized for the identification and destruction of invading pathogens, both directly and by triggering an innate immune response, *Leishmania* have evolved a number of mechanisms for suppressing some critical macrophage activities. In this review, we discuss how various species of *Leishmania* distort the host macrophage's own signalling pathways to repress the expression of various cytokines and microbicidal molecules (nitric oxide and reactive oxygen species), and antigen presentation. In particular, we describe how MAP Kinase and JAK/STAT cascades are repressed, and intracellular Ca^{2+} and the activities of protein tyrosine phosphatases, in particular SHP-1, are elevated.

Key words: *Leishmania*, macrophage, SHP-1, PTP, STAT1, nitric oxide, MHC class II.

INTRODUCTION

Single celled parasites of the genus *Leishmania* are the causative agents of leishmaniasis, a spectral disease with symptoms ranging from cutaneous lesions to a visceral form that is fatal if untreated (Herwaldt, 1999). The current range of *Leishmania* encompasses 88 countries, primarily in the developing world, from southern Europe, through Africa and the Middle East and across Asia, and throughout South and Central America as far north as the southern USA (Desjeux, 2004). It was estimated that, in the year 2000, more than 12 million individuals were infected with the various *Leishmania* species, with an estimated 1·5 to 2 million cases per year in 88 countries (World Health Organisation, 2000). 90% of new cases are reported in Afghanistan, Brazil, Iran, Peru, Saudi Arabia and Syria (World Health Organisation, 2000).

Leishmania parasites have a two-stage life cycle. Elongated, flagellated promastigotes are found in the guts and mouthparts of insect vectors (sandflies of the genus *Phlebotomus* or *Lutzomya*, depending on the geographical location). These are introduced into a mammalian host when the sandfly takes a blood meal, whereupon the promastigotes are rapidly phagocytosed by host macrophages. They then differentiate into the amastigote form, which is more rounded and lacks a flagellum. Amastigotes persist and replicate in the macrophage phagolysosome, eventually causing the macrophage to burst. The released amastigotes quickly infect other macrophages following phagocytosis in a continuing cycle of re-infection. Infected macrophages may also be ingested by another sandfly as it feeds, causing the amastigotes to differentiate into promastigotes, so completing the life cycle. A recent report has suggested a variation of this scheme, in which, on entering the host, promastigotes are phagocytosed first by polymorphonuclear neutrophil granulocytes (PMNs). These infected PMNs are themselves phagocytosed by macrophages, and the *Leishmania* then differentiate and proliferate within the phagolysosomes of the macrophages (van Zandbergen *et al.* 2004). Whatever the details of its initial entry, it is clear that the overwhelming majority of the life of *Leishmania* within the host is spent in the phagolysosome of macrophages. Since this is an extremely hostile environment, specialized for destruction of potential pathogens and, moreover, one of the principal functions of macrophages is to initiate an immune response to invading microbes, *Leishmania* have had to evolve sophisticated mechanisms for evading and repressing normal macrophage functions. In this review, we will discuss how *Leishmania* parasites promote their own survival by manipulating the macrophage's intracellular signalling pathways to repress various cytokine-inducible processes.

INHIBITION OF MACROPHAGE FUNCTIONS AND IMMUNE RESPONSES

Once *Leishmania* is well established within the macrophage, it causes aberrant activation of a

* Correspondence: Martin Olivier, McGill University, Department of Microbiology and Immunology, Duff Medical Building, Room 610, 3775 University Street, Montréal, Québec, H3A 2B4, Canada. Tel: 514-398-5592/1302. Fax: 514-398-7052. E-mail: martin.olivier@mcgill.ca

Parasitology (2005), **130**, S27–S35. © 2005 Cambridge University Press
doi:10.1017/S0031182005008139 Printed in the United Kingdom

number of macrophage functions, particularly the secretion of chemotactic chemokines (Racoosin & Beverley, 1997; Matte & Olivier, 2002; Forget & Olivier, unpublished observations). However, in the early stages of infection, inhibition of macrophage functions that are detrimental to the parasite's immediate survival appears to be more important. This is achieved by repression of gene expression, post-translational protein modification, particularly degradation, and by activation of suppressive pathways and molecules, particularly protein tyrosine phosphatases (PTPs). The repressed functions can be grouped into three broad categories: microbicidal functions; cytokine production; and antigen presentation and effector cell activation. We will consider these in turn.

Microbicidal molecules

Macrophages produce two species of highly reactive molecule that are destructive to *Leishmania*: nitric oxide (NO) (Liew *et al.* 1990) and oxygen radicals (reactive oxygen species, ROS) (Murray, 1982). Production of ROS by macrophages is detrimental to parasite survival *in vitro* (Buchmuller-Rouiller & Mauel, 1981, 1986) and ROS-deficient mice show an initial, increased susceptibility to *L. major* infection (Murray & Nathan, 1999). *Leishmania* therefore inhibit ROS production by their host macrophages (Buchmuller-Rouiller & Mauel, 1987; Olivier, Baimbridge & Reiner, 1992; Olivier, Brownsey & Reiner, 1992). This appears to be due to abnormal Protein Kinase C (PKC) activity (Olivier, Brownsey & Reiner, 1992), resulting from interaction between the macrophage and parasite surface molecules lipophosphoglycan (LPG) and/or gp63 (Sorensen, Hey & Kharazmi, 1994; Descoteaux & Turco, 1999).

NO is produced following the induction of the inducible Nitric Oxide Synthase (iNOS, NOS2) gene, and seems to be the more important for elimination of *Leishmania* since gene knock-out of iNOS in a *Leishmania*-resistant strain of mouse causes the mouse to become susceptible to *L. major* infection (Wei *et al.* 1995; Murray & Nathan, 1999). Furthermore, macrophages derived from iNOS-deficient mice are unable to eliminate *L. major in vitro* (Wei *et al.* 1995). In comparison, mice defective in ROS production eventually control the infection (Murray & Nathan, 1999). Macrophages infected with *Leishmania* promastigotes, or treated with purified *Leishmania* surface molecules LPG (Proudfoot *et al.* 1996) or glycoinositol phospholipid (GIPL) (Proudfoot, O'Donnell & Liew, 1995) are unable to induce iNOS and generate NO in response to subsequent treatment with IFNγ and/or LPS. However, LPG appears to increase NO production when it is administered to macrophages simultaneously with IFNγ (Proudfoot, O'Donnell & Liew, 1995; Proudfoot *et al.* 1996). This, together

with our recent data showing that promastigotes are unable to inhibit NO production by macrophages that have been previously stimulated with IFNγ (Whitcombe and Olivier, unpublished data), is consistent with a model in which exposure of macrophages to IFNγ, presumably produced by neighbouring NK cells and T lymphocytes, is critical for converting them to a leishmanicidal state. Inactivating the response to IFNγ is therefore vital for the survival of *Leishmania*.

Inhibition of cytokine secretion

Leishmania infection inhibits production of cytokines, primarily those involved in the inflammatory response, e.g. the IL-1 family and TNFα, or activation of T lymphocytes, e.g. IL-12. Secretion of IL-1 in response to LPS is inhibited in macrophages following either infection with *L. donovani* or *L. major* (Reiner, 1987), or incubation with LPG (Frankenburg *et al.* 1990). Interestingly, while the inhibition of IL-1β appears to be due to a transcriptional repressor acting on the gene's promoter (Hatzigeorgiou *et al.* 1996), IL-1α mRNA, but not protein, is increased following infection with *L. major*, indicating a post-transcriptional repression (Hawn *et al.* 2002). Furthermore, increased IL-1α mRNA was not observed in macrophages derived from Myd88-deficient mice, or RAW267.4 macrophage-like cells transfected with dominant-negative Myd88 (Hawn *et al.* 2002). Since Myd88 is an important intermediate in Toll-Like Receptor (TLR)-dependent signalling, this implies a role for TLRs in manipulation of macrophage activity.

Expression of TNFα in response to LPS is also inhibited in *L. donovani*-infected macrophages (Descoteaux & Matlashewski, 1989). This may be indirect, resulting from an induction of the immunosuppressive IL-10 following infection (Bhattacharyya *et al.* 2001a).

In contrast to these *in vitro* studies, we have shown an elevated presence of IL-1β and TNFα, as well as various chemokines, following infection with *L. donovani*, and particularly *L. major*, in the mouse air-pouch model (Matte & Olivier, 2002). It is unclear at this stage whether these are produced by infected macrophages, or by other members of the heterogeneous leukocyte population that is recruited by both strains of *Leishmania* (Matte & Olivier, 2002), and whether this represents a host defence mechanism, incompletely inhibited by the parasite, and/or benefits the *Leishmania*, for example by increasing the number of macrophages present, that can subsequently be infected.

IL-12 plays a critical role in the regulation of cellular immune responses. It is essential for T lymphocyte activation and subsequent IFNγ secretion, which in turn results in macrophage activation and production of microbicidal molecules.

As such, several studies have documented its suppression by *Leishmania*. *L. major* promastigotes (Carrera *et al.* 1996), *L. mexicana* amastigotes (Weinheber *et al.* 1998) and the phosphoglycan portion of LPG (Piedrafita *et al.* 1999) have all been shown to cause a marked inhibition of macrophage IL-12 production. This inhibition has been also reported in *L. major*-infected mice (Belkaid *et al.* 1998). The intracellular signalling mechanisms responsible are not well understood at this time, but it appears to involve elevated intracellular Ca^{2+} following interaction with the macrophage complement receptor CR3 and/or Fcγ receptor (Marth & Kelsall, 1997; Sutterwala *et al.* 1997), rather than an NF-κB dependent mechanism (Piedrafita *et al.* 1999).

Major histocompatibility complex expression

Development of an effective immune response against a pathogen such as *Leishmania* requires antigen presentation by specialized phagocytes to T lymphocytes. Whereas dendritic cells are now recognized to be the more important antigen presenting cell type, macrophages also present antigens through the MHC class II system, resulting in recognition by activated T lymphocytes and a reciprocal augmentation in microbicidal activities. MHC class I-mediated antigen presentation, leading to CD8+ T cell activation, is also protective against *Leishmania* in some situations (Uzonna, Joyce & Scott, 2004), but, unlike MHC class II, is not essential for the complete clearance of *Leishmania* from the host (Locksley *et al.* 1993; Huber *et al.* 1998).

Some studies have shown that *L. donovani* infection inhibits induction of MHC class I and II by IFNγ (Reiner, Ng & McMaster, 1987; Reiner *et al.* 1988) and that inhibition of MHC class II mRNA transcription is independent of cyclic AMP (Kwan *et al.* 1992). In contrast, macrophages infected with *L. amazonensis* express normal level of MHC class II (Prina *et al.* 1993; Lang *et al.* 1994*a*), but suffer defects in antigen loading (Prina *et al.* 1993). This controversy may reveal a mechanistic difference between *Leishmania* species. Defective antigen loading has also been observed in *L. major* infected macrophages (Fruth, Solioz & Louis, 1993). Furthermore, some workers have suggested that MHC class II molecules may be sequestered in the phagolysosome (Lang *et al.* 1994*b*), and even endocytosed and degraded by *L. amazonensis* amastigotes (De Souza Leao *et al.* 1995). Similarly, Kima and collaborators showed that amastigote antigens were also sequestered, and therefore unavailable for presentation by MHC class II molecules (Kima *et al.* 1996). Another study linked presentation of the LACK antigen of *L. major* and *L. amazonensis* to the developmental stage, and thus infectiousness, of the parasite: log and stationary phase promastigotes permitted

transient presentation, whereas mature (metacyclic) permitted very little and amastigotes, no detectable presentation at all (Courret *et al.* 1999). Collectively, these studies show that MHC class II-mediated antigen presentation is inhibited during *Leishmania* infection, but there remains some controversy as to the mechanism(s) involved. Future studies will therefore be of much importance.

Antigen presentation depends upon cellular communication through co-stimulatory molecules such as B7/CD28 and CD40/CD40L. It has been demonstrated that expression of B7-1 on *L. donovani*-infected macrophages is not induced by LPS stimulation (Kaye *et al.* 1994) and that this inactivation process is prostaglandin-dependent (Saha *et al.* 1995). Indeed, inhibition of CD40/CD40L binding is important for preventing iNOS production (Soong *et al.* 1996) and macrophage microbicidal activities (Kamanaka *et al.* 1996; Soong *et al.* 1996), and indeed cure of leishmaniasis depends upon a functional CD40/CD40L interaction (Campbell *et al.* 1996; Heinzel, Rerko & Hujer, 1998). Recent findings suggest that p38-dependent signalling triggered by CD40 is altered in infected macrophages, and that this may contribute to reduced iNOS expression (Awasthi *et al.* 2003). Another study suggests that *L. chagasi* infection inhibits macrophage ICAM-1 expression in response to LPS with and without IFNγ (De Almeida, Cardoso & Barral-Netto, 2003).

ALTERATION OF HOST CELL SIGNALLING

Modulation of intracellular calcium signalling and protein kinase C

One of the first second messengers that was reported to be modulated by *Leishmania* parasites was calcium (Ca^{2+}). *L. major* and *L. donovani* infections were shown to augment the basal intracellular Ca^{2+} concentration of phagocytes (Eilam, El On & Spira, 1985; Olivier, Baimbridge & Reiner, 1992). This elevation seems to result from augmented Ca^{2+} uptake by infected cells in response to depletion of intracellular Ca^{2+} stores (Mansfield & Olivier, 2002). The parasite surface molecule LPG also seems to ligate Ca^{2+} directly (Eilam, El On & Spira, 1985) and to be involved in the rapid elevation of intracellular Ca^{2+} concentration (Mansfield & Olivier, 2002). *L. donovani*-infected cells have shown altered Ca^{2+}-dependent signalling, since cells no longer respond to the chemotactic peptide f-Met-Leu-Phe and produce reactive oxygen species (ROS) (Olivier, Baimbridge & Reiner, 1992). This may be due to a reduction in host inositol tri-phosphate (IP_3) concentration (Olivier, Baimbridge & Reiner, 1992), perhaps by the action of a parasite acid phosphatase (Das *et al.* 1986), or a Ca^{2+}-dependent activation of a host IP3 phosphatase (Olivier, 1996).

Several kinases depend upon Ca^{2+} availability to be fully active, for instance classical isoforms of PKC. However, PKC activity has been shown to be reduced in *L. donovani*-infected macrophages, favouring parasite survival (McNeely & Turco, 1987; Descoteaux, Matlashewski & Turco, 1992; Olivier, Brownsey & Reiner, 1992). It appears that promastigote LPG can interact with PKC at its diacyl glycerol (DAG) binding sites, interfering with its interactions with Ca^{2+} and phospholipids, as well as blocking the insertion of PKC into the membrane (reviewed in Descoteaux & Turco, 1999). Of interest, the reduced sensitivity of PKC to DAG was first demonstrated biochemically in cells that have been infected with *L. donovani* amastigotes (Olivier, Brownsey & Reiner, 1992). Since amastigotes lack LPG, this indicates that an alternative mechanism for inhibiting PKC activity despite elevated Ca^{2+} must operate in amastigote-infected cells. *Leishmania* has been reported to inhibit MARCKs-Related Protein, a substrate of PKC associated with the cell cytoskeleton and involved in vacuole dispersal (Kane & Mosser, 2000). Another *Leishmania* surface glycolipid, GIPL, may also inactivate PKC, but the underlying mechanisms are unclear (McNeely *et al.* 1989). More recently, it has been demonstrated that *Leishmania*-mediated PKC inhibition applies only to the family members that are dependent upon Ca^{2+} for their full activation (Bhattacharyya *et al.* 2001*b*). This selective inhibition may be due to the negative action of ceramides and IL-10 produced by infected phagocytes (Bhattacharyya *et al.* 2001*a,b*; Ghosh *et al.* 2001).

Inhibition of JAK/STAT pathways

Given that some of the important macrophage functions suppressed by *Leishmania* are IFNγ-inducible (e.g. NO production, MHC class II expression), it is not surprising that *Leishmania* can inhibit the JAK2/STAT1 signalling cascade, which is known to be the major pathway activated by IFNγ receptor activation. In fact, infected macrophages show defective phosphorylation of JAK1, JAK2 and STAT1 on IFNγ stimulation (Nandan & Reiner, 1995; Blanchette *et al.* 1999). This inactivation depends on the activation of phosphotyrosine phosphatases (PTPs), in particular the PTP, SHP-1. One study has shown that inactivation of JAK2/STAT1 is caused by the negative regulation of the IFNγ receptor in infected cells (Ray *et al.* 2000). However, this conflicts with the observations that other IFNγ-dependent events, such as MHC class I-mediated antigen presentation, are not affected by infection (Kima, Ruddle & McMahon-Pratt, 1997).

In a recent study, we investigated the mechanisms by which infection with *L. donovani* inactivates macrophage STAT1α. We observed that the reduced activity of STAT1α in the nucleus of infected cells is caused by rapid and sustained STAT1α protein degradation. This phenomenon seems to be specific to STAT1α, since STAT3 levels were unchanged. Using both PTP inhibitors and SHP-1 deficient macrophages, we found that this was a PTP independent process. Instead, we found that the phenomenon was reversed by inhibitors of both the proteasome and PKC. Taken together, these results argue for a direct role of the proteasome pathway in the specific proteolysis of STAT1α in macrophages infected with *Leishmania*, representing a new mechanism whereby pathogens can subvert microbicidal actions.

Inhibition of MAP kinase cascades

Infection of naïve macrophages by *L. donovani* has been shown not to activate the most important mitogen activated proteins kinase (MAPK) family members: Extracellular-Regulated Kinase (ERK1/2), p38, and Jun N-terminal Kinase (JNK) (Prive & Descoteaux, 2000), consistent with a 'silent' model of entry. A number of other studies, however, go further and show that MAPKs are actively repressed, and cannot be activated when infected macrophages are stimulated with a variety of agonists. In one study, ceramide-mediated inactivation of ERK1/2, and a resulting inhibition of the transcription factors AP-1 and NF-κB, was proposed to explain the absence of NO generation by *L. donovani*-infected cells (Ghosh *et al.* 2002). Most studies, however, have focused on suppression of MAPK cascades by dephosphorylation. In one case, *L. amazonensis* amastigotes were shown to rapidly alter phosphorylation of ERK1 in response to LPS, and this was linked to the activity of a parasite Protein Tyrosine Phosphatase (PTP) (Martiny *et al.* 1999). Another suggested that a host PTP was responsible for the inactivation of ERK1/2, and resulted in the repression of Elk-1 and c-fos transcription factor expression (Nandan, Lo & Reiner, 1999). This aspect is discussed in greater detail below.

Other MAP kinase family members are also inactivated by infection. For example, the p38 MAPK has shown abnormal activity in *L. major*-infected cells that are stimulated with a CD40 activating antibody, which mimics the macrophage-T lymphocyte interaction (Awasthi *et al.* 2003). Consistent with previous work, infected macrophages in this study cannot express iNOS: the authors therefore suggest that this may, in part, result from p38 inactivation. This is supported by the observation that treatment of macrophages with anisomycin, an activator of p38, reduces the survival of *L. donovani* (Junghae & Raynes, 2002). Such conclusions must be treated with caution, however, since the use of any compounds that contribute to macrophage activation might be expected to hinder parasite survival.

The parasite surface molecule LPG has been implicated in the inactivation of MAPKs, since phagocytosis of LPG deficient *L. donovani* promastigotes caused MAPK activation, without the requirement for subsequent macrophage stimulation (Prive & Descoteaux, 2000). This does not necessarily mean that LPG itself directly causes inhibition but that, in the absence of LPG, the interaction of *Leishmania* with its host cell, and thus the stimulation of host cell receptors, is modified. We have observed that LPG-deficient *L. donovani* still induce the PTP SHP-1 (see below), with resulting inactivation of JAK2 and MAPK signalling pathways and downstream transcription factors (Olivier *et al.*, unpublished data).

Tyrosine dephosphorylation by SHP-1

As discussed above, inhibition of MAPK signalling pathways is due, at least in a large part, to the activity of Protein Tyrosine Phosphatases (PTPs) (Martiny *et al.* 1999; Nandan, Lo & Reiner, 1999). We recently obtained clear evidence that activation of the host PTP SHP-1 is responsible for the dephosphorylation and inactivation of ERK1/2, as SHP-1-deficient macrophages showed normal JAK2 and ERK1/2 activity following infection with *L. donovani*, and responded to IFNγ by increased NO production (Forget *et al.*, unpublished). In a parallel study, we have demonstrated that a general tyrosine dephosphorylation of high molecular weight proteins occurred rapidly – even within 5 minutes – following the initial interaction of macrophages with *L. donovani* promastigotes (Blanchette *et al.* 1999), and amastigotes (Abu-Dayyeh & Olivier, unpublished). This was accompanied by an increase of PTP activity, in particular the activity of SHP-1 (Blanchette *et al.* 1999). We also observed that SHP-1 directly associated with JAK2 upon *L. donovani* infection and this interaction may well account for the absence of JAK2 phosphorylation, and thus activation, following IFNγ stimulation (Blanchette *et al.* 1999). More recently, using a bone marrow-derived macrophage cell line deficient in SHP-1, we have firmly established that this PTP is responsible for the inactivation of JAK2 and ERK1/2 MAPK. The restoration of these signalling pathways in SHP-1-deficient cells also restores the ability of infected cells to generate NO in response to IFNγ stimulation (Forget *et al.*, unpublished). Our studies are supported by others, who have reported that SHP-1 strongly interacts with MAPKs following *L. donovani* infection (Nandan, Lo & Reiner, 1999). In this study, infected macrophages did not respond to PMA, an artificial PKC activator, and this was reflected by an inhibition of ERK1/2 phosphorylation, Elk-1 activation and c-fos mRNA expression. A striking difference between this latter study and ours is that the inhibition of ERK1/2 was observed

17 hours post-infection, whereas we and others observed SHP-1 activation and JAK2 inhibition within the first hour post-infection (Blanchette *et al.* 1999; Martiny *et al.* 1999; Forget *et al.*, unpublished). Differences between cell lines and their status of differentiation may contribute towards this discrepancy. However, the rapid inhibition of host cell functions is vital for parasite survival. Recently, it has been claimed that the elongation factor EF-1α of *Leishmania* could be responsible for the late induction of SHP-1 observed 16 hours post-infection (Nandan *et al.* 2002; Nandan & Reiner, 2005). However, the authors did not prove the claim by generating an EF-1α deficient parasite. In our opinion, the rapidity of SHP-1 activation requires a receptor-mediated process, although another mechanism may contribute at later time-points.

The biological significance of these studies is emphasised by our *in vivo* studies, using mice deficient for SHP-1, or treated with peroxovanadium compounds, which are potent PTP inhibitors. Treatment with peroxovanadium compounds almost completely protects against development of murine models of cutaneous and visceral leishmaniasis, following inoculation with *L. major* and *L. donovani*, respectively (Matte *et al.* 2000). Interestingly, this protection is not observed in iNOS-deficient mice (Matte *et al.* 2000). This is consistent with the *in vitro* studies described above which suggested that the activation of SHP-1 by *Leishmania* is important for the prevention of NO generation. Mice of the viable motheaten strain, which are deficient for SHP-1, also do not develop swelling following infection of footpads with *L. major*, and have a significantly reduced parasite load (Forget *et al.* 2001). Histochemical analysis and *in situ* hybridization of the infected footpads showed that, unlike wild-type animals, iNOS gene expression and the activities of STAT1 and NF-κB, which are know to contribute to iNOS induction, was not inhibited following infection. In fact, there was an activation of these intermediates, contributing to increased iNOS expression and thus NO generation (Forget *et al.* 2001). This was supported by our finding that aminoguanidine, an inhibitor of iNOS, completely reversed the protection against *L. major* observed in SHP-1 deficient animals (Forget *et al.* 2001). We recently obtained similar observations concerning the role of SHP-1 in the control of visceral leishmaniasis conferred by *L. donovani* infection *in vivo*. Interestingly, we found by a pathogenomic approach that repression of genes involved in intracellular signalling and the immune response by *L. donovani* is almost entirely SHP-1 dependent (Olivier *et al.*, unpublished).

CONCLUDING REMARKS

The studies we have discussed here described some of the mechanisms employed by *Leishmania* to inhibit important macrophage functions by distorting the host cell's own signalling pathways, and clearly show the importance of this ability for parasite survival. While the vital roles played by signalling intermediates such as SHP-1 are apparent, the up-stream events that initiate these processes are still unclear. In particular, describing the mechanisms of interaction between the parasite and receptors on the macrophage surface that lead to activation of the suppressive pathways will be of great importance in the next few years. Since strategies for therapeutic manipulation of signalling intermediates exist, we hope that the increasing understanding of the signalling strategies employed by *Leishmania* will soon have significance for the treatment of leishmaniasis. Finally, the insights into macrophage inactivation and immune evasion gained from these studies have relevance for a wide range of both parasitic infections and autoimmune situations.

ACKNOWLEDGMENTS

M.O. is supported by grants from the Canadian Institutes in Health Research (CIHR) and is member of a CIHR Group in Host-Pathogen Interaction. M.O. is recipient of a Fonds de la Recherche en Santé du Québec (FRSQ) Senior Scholarship and is a Burroughs Wellcome Fund Awardee in Molecular Parasitology. D.J.G. is a recipient of a post-doctoral training award from the CIHR Centre for Host Resistance at the Research Institute of McGill University Health Centre.

REFERENCES

AWASTHI, A., MATHUR, R., KHAN, A., JOSHI, B. N., JAIN, N., SAWANT, S., BOPPANA, R., MITRA, D. & SAHA, B. (2003). CD40 signaling is impaired in *L. major*-infected macrophages and is rescued by a p38MAPK activator establishing a host-protective memory T cell response. *Journal of Experimental Medicine* **197**, 1037–1043.

BELKAID, Y., BUTCHER, B. & SACKS, D. L. (1998). Analysis of cytokine production by inflammatory mouse macrophages at the single-cell level: selective impairment of IL-12 induction in *Leishmania*-infected cells. *European Journal of Immunology* **28**, 1389–1400.

BHATTACHARYYA, S., GHOSH, S., JHONSON, P. L., BHATTACHARYA, S. K. & MAJUMDAR, S. (2001a). Immunomodulatory role of interleukin-10 in visceral leishmaniasis: defective activation of protein kinase C-mediated signal transduction events. *Infection and Immunity* **69**, 1499–1507.

BHATTACHARYYA, S., GHOSH, S., SEN, P., ROY, S. & MAJUMDAR, S. (2001b). Selective impairment of protein kinase C isotypes in murine macrophage by *Leishmania donovani*. *Molecular and Cellular Biochemistry* **216**, 47–57.

BLANCHETTE, J., RACETTE, N., FAURE, R., SIMINOVITCH, K. A. & OLIVIER, M. (1999). *Leishmania*-induced increases in activation of macrophage SHP-1 tyrosine phosphatase are associated with impaired IFN-gamma-triggered JAK2 activation. *European Journal of Immunology* **29**, 3737–3744.

BUCHMULLER-ROUILLER, Y. & MAUEL, J. (1981). Studies on the mechanisms of macrophage activation – possible involvement of oxygen metabolites in killing of *Leishmania*-Enriettii by activated mouse macrophages. *Journal of the Reticuloendothelial Society* **29**, 181–192.

BUCHMULLER-ROUILLER, Y. & MAUEL, J. (1986). Correlation between enhanced oxidative metabolism and leishmanicidal activity in activated macrophages from healer and nonhealer mouse strains. *Journal of Immunology* **136**, 3884–3890.

BUCHMULLER-ROUILLER, Y. & MAUEL, J. (1987). Impairment of macrophage oxidative-metabolism by *Leishmania*. *Experientia* **43**, 665–665.

CAMPBELL, K. A., OVENDALE, P. J., KENNEDY, M. K., FANSLOW, W. C., REED, S. G. & MALISZEWSKI, C. R. (1996). CD40 ligand is required for protective cell-mediated immunity to *Leishmania major*. *Immunity* **4**, 283–289.

CARRERA, L., GAZZINELLI, R. T., BADOLATO, R., HIENY, S., MULLER, W., KUHN, R. & SACKS, D. L. (1996). *Leishmania* promastigotes selectively inhibit interleukin 12 induction in bone marrow-derived macrophages from susceptible and resistant mice. *Journal of Experimental Medicine* **183**, 515–526.

COURRET, N., PRINA, E., MOUGNEAU, E., SARAIVA, E. M., SACKS, D. L., GLAICHENHAUS, N. & ANTOINE, J. C. (1999). Presentation of the *Leishmania* antigen LACK by infected macrophages is dependent upon the virulence of the phagocytosed parasites. *European Journal of Immunology* **29**, 762–773.

DAS, S., SAHA, A. K., REMALEY, A. T., GLEW, R. H., DOWLING, J. N., KAJIYOSHI, M. & GOTTLIEB, M. (1986). Hydrolysis of phosphoproteins and inositol phosphates by cell surface phosphatase of *Leishmania donovani*. *Molecular and Biochemical Parasitology* **20**, 143–153.

DE ALMEIDA, M. C., CARDOSO, S. A. & BARRAL-NETTO, M. (2003). *Leishmania (Leishmania) chagasi* infection alters the expression of cell adhesion and costimulatory molecules on human monocyte and macrophage. *International Journal for Parasitology* **33**, 153–162.

DE SOUZA LEAO S., LANG, T., PRINA, E., HELLIO, R. & ANTOINE, J. C. (1995). Intracellular *Leishmania amazonensis* amastigotes internalize and degrade MHC class II molecules of their host cells. *Journal of Cell Science* **108**, 3219–3231.

DESCOTEAUX, A. & MATLASHEWSKI, G. (1989). c-fos and tumor necrosis factor gene expression in *Leishmania donovani*-infected macrophages. *Molecular Cell Biology* **9**, 5223–5227.

DESCOTEAUX, A., MATLASHEWSKI, G. & TURCO, S. J. (1992). Inhibition of macrophage protein kinase C-mediated protein phosphorylation by *Leishmania donovani* lipophosphoglycan. *Journal of Immunology* **149**, 3008–3015.

DESCOTEAUX, A. & TURCO, S. J. (1999). Glycoconjugates in *Leishmania* infectivity. *Biochimica et Biophysica Acta* **1455**, 341–352.

DESJEUX, P. (2004). Leishmaniasis. *Nature Reviews Microbiology* **2**, 692.

EILAM, Y., EL ON, J. & SPIRA, D. T. (1985). *Leishmania major*: excreted factor, calcium ions, and the survival of amastigotes. *Experimental Parasitology* **59**, 161–168.

FORGET, G., SIMINOVITCH, K. A., BROCHU, S., RIVEST, S., RADZIOCH, D. & OLIVIER, M. (2001). Role of host phosphotyrosine phosphatase SHP-1 in the development of murine leishmaniasis. *European Journal of Immunology* **31**, 3185–3196.

FRANKENBURG, S., LEIBOVICI, V., MANSBACH, N., TURCO, S. J. & ROSEN, G. (1990). Effect of glycolipids of *Leishmania* parasites on human monocyte activity. Inhibition by lipophosphoglycan. *Journal of Immunology* **145**, 4284–4289.

FRUTH, U., SOLIOZ, N. & LOUIS, J. A. (1993). *Leishmania major* interferes with antigen presentation by infected macrophages. *Journal of Immunology* **150**, 1857–1864.

GHOSH, S., BHATTACHARYYA, S., DAS, S., RAHA, S., MAULIK, N., DAS, D. K., ROY, S. & MAJUMDAR, S. (2001). Generation of ceramide in murine macrophages infected with *Leishmania donovani* alters macrophage signaling events and aids intracellular parasitic survival. *Molecular and Cellular Biochemistry* **223**, 47–60.

GHOSH, S., BHATTACHARYYA, S., SIRKAR, M., SA, G. S., DAS, T., MAJUMDAR, D., ROY, S. & MAJUMDAR, S. (2002). *Leishmania donovani* suppresses activated protein 1 and NF-kappaB activation in host macrophages via ceramide generation: involvement of extracellular signal-regulated kinase. *Infection and Immunity* **70**, 6828–6838.

HATZIGEORGIOU, D. E., GENG, J., ZHU, B., ZHANG, Y., LIU, K., ROM, W. N., FENTON, M. J., TURCO, S. J. & HO, J. L. (1996). Lipophosphoglycan from *Leishmania* suppresses agonist-induced interleukin 1 beta gene expression in human monocytes via a unique promoter sequence. *Proceedings of the National Academy of Sciences, USA* **93**, 14708–14713.

HAWN, T. R., OZINSKY, A., UNDERHILL, D. M., BUCKNER, F. S., AKIRA, S. & ADEREM, A. (2002). *Leishmania major* activates IL-1 alpha expression in macrophages through a MyD88-dependent pathway. *Microbes and Infection* **4**, 763–771.

HEINZEL, F. P., RERKO, R. M. & HUJER, A. M. (1998). Underproduction of interleukin-12 in susceptible mice during progressive leishmaniasis is due to decreased CD40 activity. *Cellular Immunology* **184**, 129–142.

HERWALDT, B. L. (1999). Leishmaniasis. *Lancet* **354**, 1191–1199.

HUBER, M., TIMMS, E., MAK, T. W., ROLLINGHOFF, M. & LOHOFF, M. (1998). Effective and long-lasting immunity against the parasite *Leishmania major* in CD8-deficient mice. *Infection and Immunity* **66**, 3968–3970.

JUNGHAE, M. & RAYNES, J. G. (2002). Activation of p38 mitogen-activated protein kinase attenuates *Leishmania donovani* infection in macrophages. *Infection and Immunity* **70**, 5026–5035.

KAMANAKA, M., YU, P., YASUI, T., YOSHIDA, K., KAWABE, T., HORII, T., KISHIMOTO, T. & KIKUTANI, H. (1996). Protective role of CD40 in *Leishmania major* infection at two distinct phases of cell-mediated immunity. *Immunity* **4**, 275–281.

KANE, M. M. & MOSSER, D. M. (2000). *Leishmania* parasites and their ploys to disrupt macrophage activation. *Current Opinion in Hematology* **7**, 26–31.

KAYE, P. M., ROGERS, N. J., CURRY, A. J. & SCOTT, J. C. (1994). Deficient expression of co-stimulatory molecules on *Leishmania*-infected macrophages. *European Journal of Immunology* **24**, 2850–2854.

KIMA, P. E., RUDDLE, N. H. & McMAHON-PRATT, D. (1997). Presentation via the class I pathway by *Leishmania amazonensis*-infected macrophages of an endogenous

leishmanial antigen to CD8+ T cells. *Journal of Immunology* **159**, 1828–1834.

KIMA, P. E., SOONG, L., CHICHARRO, C., RUDDLE, N. H. & McMAHON-PRATT, D. (1996). *Leishmania*-infected macrophages sequester endogenously synthesized parasite antigens from presentation to CD4+ T cells. *European Journal of Immunology* **26**, 3163–3169.

KWAN, W. C., McMASTER, W. R., WONG, N. & REINER, N. E. (1992). Inhibition of expression of major histocompatibility complex class II molecules in macrophages infected with *Leishmania donovani* occurs at the level of gene transcription via a cyclic AMP-independent mechanism. *Infection and Immunity* **60**, 2115–2120.

LANG, T., DE CHASTELLIER, C., FREHEL, C., HELLIO, R., METEZEAU, P., LEAO, S. S. & ANTOINE, J. C. (1994a). Distribution of MHC class I and of MHC class II molecules in macrophages infected with *Leishmania amazonensis*. *Journal of Cell Science* **107**, 69–82.

LANG, T., HELLIO, R., KAYE, P. M. & ANTOINE, J. C. (1994b). *Leishmania donovani*-infected macrophages: characterization of the parasitophorous vacuole and potential role of this organelle in antigen presentation. *Journal of Cell Science* **107**, 2137–2150.

LIEW, F. Y., MILLOTT, S., PARKINSON, C., PALMER, R. M. & MONCADA, S. (1990). Macrophage killing of *Leishmania* parasite *in vivo* is mediated by nitric oxide from L-arginine. *Journal of Immunology* **144**, 4794–4797.

LOCKSLEY, R. M., REINER, S. L., HATAM, F., LITTMAN, D. R. & KILLEEN, N. (1993). Helper T cells without CD4: control of leishmaniasis in CD4-deficient mice. *Science* **261**, 1448–1451.

MANSFIELD, J. M. & OLIVIER, M. (2002). Immune Evasion by Parasites. In *Immunology of Infectious Diseases* (ed. Kaufmann, S. H. E., Sher, A. & Ahmed, R.), pp. 379–392. ASM Press, Washington, DC.

MARTH, T. & KELSALL, B. L. (1997). Regulation of interleukin-12 by complement receptor 3 signaling. *Journal of Experimental Medicine* **185**, 1987–1995.

MARTINY, A., MEYER-FERNANDES, J. R., DE SOUZA, W. & VANNIER-SANTOS, M. A. (1999). Altered tyrosine phosphorylation of ERK1 MAP kinase and other macrophage molecules caused by *Leishmania* amastigotes. *Molecular and Biochemical Parasitology* **102**, 1–12.

MATTE, C., MARQUIS, J. F., BLANCHETTE, J., GROS, P., FAURE, R., POSNER, B. I. & OLIVIER, M. (2000). Peroxovanadium-mediated protection against murine leishmaniasis: role of the modulation of nitric oxide. *European Journal of Immunology* **30**, 2555–2564.

MATTE, C. & OLIVIER, M. (2002). Leishmania-induced cellular recruitment during the early inflammatory response: modulation of proinflammatory mediators. *Journal of Infectious Disease* **185**, 673–681.

McNEELY, T. B., ROSEN, G., LONDNER, M. V. & TURCO, S. J. (1989). Inhibitory effects on protein kinase C activity by lipophosphoglycan fragments and glycosylphosphatidylinositol antigens of the protozoan parasite *Leishmania*. *Biochemical Journal* **259**, 601–604.

McNEELY, T. B. & TURCO, S. J. (1987). Inhibition of protein kinase C activity by the *Leishmania donovani* lipophosphoglycan. *Biochemical and Biophysical Research Communications* **148**, 653–657.

MURRAY, H. W. (1982). Cell-mediated immune response in experimental visceral leishmaniasis. II. Oxygen-dependent killing of intracellular *Leishmania donovani* amastigotes. *Journal of Immunology* **129**, 351–357.

MURRAY, H. W. & NATHAN, C. F. (1999). Macrophage microbicidal mechanisms *in vivo*: reactive nitrogen versus oxygen intermediates in the killing of intracellular visceral *Leishmania donovani*. *Journal of Experimental Medicine* **189**, 741–746.

NANDAN, D., LO, R. & REINER, N. E. (1999). Activation of phosphotyrosine phosphatase activity attenuates mitogen-activated protein kinase signaling and inhibits c-FOS and nitric oxide synthase expression in macrophages infected with *Leishmania donovani*. *Infection and Immunity* **67**, 4055–4063.

NANDAN, D. & REINER, N. E. (1995). Attenuation of gamma-interferon-induced tyrosine phosphorylation in mononuclear phagocytes infected with *Leishmania donovani* – selective-inhibition of signaling through Janus Kinases and Stat1. *Infection and Immunity* **63**, 4495–4500.

NANDAN, D. & REINER, N. E. (2005). *Leishmania donovani* engages in regulatory interference by targeting macrophage protein tyrosine phosphatase SHP-1. *Clinical Immunology* **114**, 266–277.

NANDAN, D., YI, T. L., LOPEZ, M., LAI, C. & REINER, N. E. (2002). *Leishmania* EF-1 alpha activates the Src homology 2 domain containing tyrosine phosphatase SHP-1 leading to macrophage deactivation. *Journal of Biological Chemistry* **277**, 50190–50197.

OLIVIER, M. (1996). Modulation of host cell intracellular Ca^{2+}. *Parasitology Today* **12**, 145–150.

OLIVIER, M., BAIMBRIDGE, K. G. & REINER, N. E. (1992). Stimulus-response coupling in monocytes infected with *Leishmania* – attenuation of calcium transients is related to defective agonist-induced accumulation of inositol phosphates. *Journal of Immunology* **148**, 1188–1196.

OLIVIER, M., BROWNSEY, R. W. & REINER, N. E. (1992). Defective stimulus-response coupling in human monocytes infected with *Leishmania donovani* is associated with altered activation and translocation of protein-kinase-C. *Proceedings of the National Academy of Sciences, USA* **89**, 7481–7485.

PIEDRAFITA, D., PROUDFOOT, L., NIKOLAEV, A. V., XU, D. M., SANDS, W., FENG, G. J., THOMAS, E., BREWER, J., FERGUSON, M. A. J., ALEXANDER, J. & LIEW, F. Y. (1999). Regulation of macrophage IL-12 synthesis by *Leishmania* phosphoglycans. *European Journal of Immunology* **29**, 235–244.

PRINA, E., JOUANNE, C., DE SOUZA, L. S., SZABO, A., GUILLET, J. G. & ANTOINE, J. C. (1993). Antigen presentation capacity of murine macrophages infected with *Leishmania amazonensis* amastigotes. *Journal of Immunology* **151**, 2050–2061.

PRIVE, C. & DESCOTEAUX, A. (2000). *Leishmania donovani* promastigotes evade the activation of mitogen-activated protein kinases p38, c-Jun N-terminal kinase, and extracellular signal-regulated kinase-1/2 during infection of naive macrophages. *European Journal of Immunology* **30**, 2235–2244.

PROUDFOOT, L., NIKOLAEV, A. V., FENG, G. J., WEI, W. Q., FERGUSON, M. A., BRIMACOMBE, J. S. & LIEW, F. Y. (1996). Regulation of the expression of nitric oxide synthase and leishmanicidal activity by glycoconjugates of *Leishmania* lipophosphoglycan in murine macrophages. *Proceedings of the National Academy of Sciences, USA* **93**, 10984–10989.

PROUDFOOT, L., O'DONNELL, C. A. & LIEW, F. Y. (1995). Glycoinositolphospholipids of *Leishmania major* inhibit nitric oxide synthesis and reduce leishmanicidal activity in murine macrophages. *European Journal of Immunology* **25**, 745–750.

RACOOSIN, E. L. & BEVERLEY, S. M. (1997). *Leishmania major*: promastigotes induce expression of a subset of chemokine genes in murine macrophages. *Experimental Parasitology* **85**, 283–295.

RAY, M., GAM, A. A., BOYKINS, R. A. & KENNEY, R. T. (2000). Inhibition of interferon-gamma signaling by *Leishmania donovani*. *Journal of Infectious Diseases* **181**, 1121–1128.

REINER, N. E. (1987). Parasite accessory cell-interactions in murine leishmaniasis.1. Evasion and stimulus-dependent suppression of the macrophage interleukin-1 response by *Leishmania-donovani*. *Journal of Immunology* **138**, 1919–1925.

REINER, N. E., NG, W., MA, T. & McMASTER, W. R. (1988). Macrophages infected with *Leishmania-donovani* have suppressed responses to interferon-gamma for the induction of major histocompatibility complex class-Ii messenger-RNA. *Clinical Research* **36**, A468–A468.

REINER, N. E., NG, W. & McMASTER, W. R. (1987). Parasite-accessory cell-interactions in murine leishmaniasis 2. *Leishmania donovani* suppresses macrophage expression of class-I and class-Ii major histocompatibility complex gene-products. *Journal of Immunology* **138**, 1926–1932.

SAHA, B., DAS, G., VOHRA, H., GANGULY, N. K. & MISHRA, G. C. (1995). Macrophage-T cell interaction in experimental visceral leishmaniasis: failure to express costimulatory molecules on *Leishmania*-infected macrophages and its implication in the suppression of cell-mediated immunity. *European Journal of Immunology* **25**, 2492–2498.

SOONG, L., XU, J. C., GREWAL, I. S., KIMA, P., SUN, J., LONGLEY, B. J., RUDDLE, N. H., McMAHON-PRATT, D. & FLAVELL, R. A. (1996). Disruption of CD40-CD40 ligand interactions results in an enhanced susceptibility to *Leishmania amazonensis* infection. *Immunity* **4**, 263–273.

SORENSEN, A. L., HEY, A. S. & KHARAZMI, A. (1994). *Leishmania major* surface protease Gp63 interferes with the function of human monocytes and neutrophils *in vitro*. *Acta Pathologica, Microbiologica et Immunologica Scandinavica* **102**, 265–271.

SUTTERWALA, F. S., NOEL, G. J., CLYNES, R. & MOSSER, D. M. (1997). Selective suppression of interleukin-12 induction after macrophage receptor ligation. *Journal of Experimental Medicine* **185**, 1977–1985.

UZONNA, J. E., JOYCE, K. L. & SCOTT, P. (2004). Low dose *Leishmania major* promotes a transient T helper cell type 2 response that is down-regulated by interferon gamma-producing CD8+ T Cells. *Journal of Experimental Medicine* **199**, 1559–1566.

VAN ZANDBERGEN, G., KLINGER, M., MUELLER, A., DANNENBERG, S., GEBERT, A., SOLBACH, W. & LASKAY, T.

(2004). Cutting edge: neutrophil granulocyte serves as a vector for *Leishmania* entry into macrophages. *Journal of Immunology* **173**, 6521–6525.

WEI, X. Q., CHARLES, I. G., SMITH, A., URE, J., FENG, G. J., HUANG, F. P., XU, D., MULLER, W., MONCADA, S. & LIEW, F. Y. (1995). Altered immune responses in mice lacking inducible nitric oxide synthase. *Nature* **375**, 408–411.

WEINHEBER, N., WOLFRAM, M., HARBECKE, D. & AEBISCHER, T. (1998). Phagocytosis of *Leishmania mexicana* amastigotes by macrophages leads to a sustained suppression of IL-12 production. *European Journal of Immunology* **28**, 2467–2477.

WORLD HEALTH ORGANISATION (2000). *The Leishmaniases and Leishmania/HIV co-infections*. WHO Factsheet No. 116. WHO, Geneva.

Constitutively activated CK2 potentially plays a pivotal role in *Theileria*-induced lymphocyte transformation

F. DESSAUGE[1,2], R. LIZUNDIA[1] *and* G. LANGSLEY[1]*

[1] *Laboratoire de Biologie Cellulaire Comparative des Apicomplexes, UMR 8104 CNRS/U567 INSERM, Département Maladies Infectieuses, Hôpital Cochin – Bâtiment Gustave Roussy, Institut Cochin, 27, rue du Faubourg-Saint-Jacques, 75014 Paris, France*
[2] *Laboratoire d'Immunologie Cellulaire et Tissulaire, INSERM U543, Bâtiment CERVI Groupe Hospitalier Pitié-Salpêtrière, 83, Boulevard de l'Hôpital, 75651 Paris Cedex 13 France*

SUMMARY

Activation of casein kinase II (CK2) was one of the first observations made on how *Theileria* parasites manipulate host cell signal transduction pathways and we argue that CK2 induction may in fact contribute to many of the different activation events that have been described since 1993 for *Theileria*-infected lymphocytes such as sustained activation of transcription factors c-Myc and NF-κB. CK2 also contributes to infected lymphocyte survival by inhibiting caspase activation and is probably behind constitutive PI3-K activation by phosphorylating PTEN. Finally, we also discuss how CK2A may act not only as a kinase, but also as a stimulatory subunit for the protein phosphatase PP2A, so dampening down the MEK/ERK and Akt/PKB pathways and for all these reasons we propose CK2 as a central player in *Theileria*-induced lymphocyte transformation.

Key words: *Theileria*, transformation, CK2, Myc, apoptosis.

INTRODUCTION

Theileria parva is an obligate, intracellular, parasitic protozoan that causes East Coast fever, an acute leukaemia-like disease of cattle. *T. parva* and *T. annulata* (causative agent of Tropical Theileriosis) are unique among eukaryotes in that infection induces transformation of their host cells, namely, bovine leukocytes. One of the earliest observations on host cell activation induced by infection showed that *T. parva*-infected lymphocytes display a 4- to 12-fold increase in total protein phosphorylation and that this is largely due to a serine/threonine kinase called casein kinase II (CK2, (ole-MoiYoi *et al.* 1993)). Augmented kinase activity correlates with increased transcription, as there is a 4- to 6-fold increase in *CK2a* mRNA in the infected cells relative to controls and a marked increase in the amount of CK2 protein in infected cell lines. Bovine lymphocyte CK2 therefore, appears to be constitutively activated in *Theileria*-infected cells. *Theileria* parasites however, code for their own CK2 (ole-MoiYoi *et al.* 1992; Shayan & Ahmed, 1997), but the enzyme does not appear to be secreted into the host cell cytoplasm and it is unlikely therefore, that parasite kinase activity underlies the increased mammalian CK2 activity associated with infected leukocytes (Biermann *et al.* 2003).

CK2 is a heterodimeric enzyme made up of two catalytic alpha subunits in a complex with two regulatory beta subunits (Fig. 1). The complex is dynamic and the different subunits can associate with other complexes (Filhol, Martiel & Cochet 2004). Analysis of the human gene for *CK2α* showed promoter activity resides essentially in a region between positions -9 to 46 and binding sites for the transcription factors NF-κB and Sp1 were identified (Krehan *et al.* 2000). Interestingly, NF-κB is strongly activated by *Theileria*-infection (Heussler *et al.* 1999), so perhaps parasite-mediated activation of NF-κB might be responsible for the increased transcription of *CK2α* (ole-MoiYoi *et al.* 1993). However, cytokines such as TNF-α and growth factors can also stimulate CK2 activity (Sayed *et al.* 2000) and *T. parva*-infected B cells proliferate in part due to a TNF autocrine loop (Guergnon *et al.* 2003*a*), implying that TNF (and potentially other cytokines) might contribute to the sustained CK2 activity of infected cells. Whatever the origins of CK2 induction in *Theileria*-infected leukocytes, the potential consequences of sustained CK2A over-expression were elegantly demonstrated in a transgenic mouse model (Seldin & Leder, 1995) and the similarity with *Theileria*-provoked lymphoproliferation has been remarked upon (ole-MoiYoi, 1995). What was striking was that adult transgenic mice, like *Theileria*-infected lymphocytes, displayed a propensity to develop lymphomas, but when this occurred in the context of *c-myc* over-expression the mice developed neonatal leukemia. This is particularly striking, since *Theileria*-infected B cells have just been described as having constitutively high

* Corresponding author. E-mail: langsley@cochin.inserm.fr

Parasitology (2005), **130**, S37–S44. © 2005 Cambridge University Press
doi:10.1017/S0031182005008140 Printed in the United Kingdom

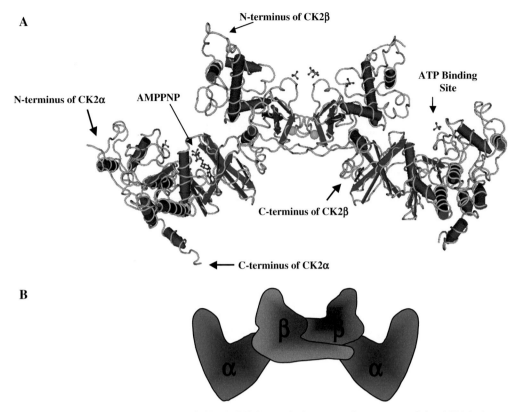

Fig. 1. Structure of the tetrameric CK2. A. High-resolution crystal structure of the CK2 holoenzyme (Niefind *et al.* 2001) was obtained with Cn3D program to generate a ribbon diagram illustrating the CK2 tetramer. The high-resolution structure of tetrameric CK2 demonstrated that the non-hydrolysable ATP analogue adenosine 5′-[β,γ-imido]triphosphate (AMPPNP) is present in the ATP binding site of only one of the catalytic CK2A subunits within the CK2 tetramer. Anti-parallel β strands are illustrated in blue and alpha helix strands are illustrated in red. B). Schematic structure of tetrameric CK2 with the two catalytic CK2A subunits and the two regulatory CK2B subunits.

levels of c-Myc (Dessauge *et al.* 2005). Thus, the co-induction of CK2 and c-Myc may be the major driving force behind *Theileria*-induced lymphocyte transformation.

Here, we argue that in addition to synergy with c-Myc, CK2 could influence a number of other host cell pathways induced by *Theileria*, such as activation of the lipid inositol 3-kinase (Baumgartner *et al.* 2000; Heussler *et al.* 2001) and we illustrate this with some experimental evidence that supports a pivotal role for CK2 in *Theileria*-induced lymphocyte transformation.

CK2 AUGMENTS C-MYC-DRIVEN TRANSCRIPTION IN *T. PARVA*-INFECTED B CELLS

As stated, transgenic mice that over-express *CK2a* develop lymphomas particularly when co-expressed with a *c-myc* transgene and pharmacologic inhibition of CK2 activity in the T cell lymphomas reduced both their proliferation and the levels of c-Myc protein (Channavajhala & Seldin, 2002). This is due to proteasome-dependent accelerated turnover of c-Myc protein and inhibition of *CK2* transcription with anti-sense *CK2* constructs modulated c-Myc protein levels (Channavajhala & Seldin, 2002).

Similarly, in *T. parva*-transformed bovine B cells pharmacological inhibition of CK2 with apigenin also reduced c-Myc activity (Dessauge *et al.* 2005) and over-expression of a kinase-dead mutant of CK2 also reduces *c-myc*-driven transcription (Fig. 2). In contrast, over-expression of active wild-type CK2 kinase stabilizes c-Myc and results in enhanced c-Myc-driven luciferase activity (Fig. 2). Thus in *Theileria*-infected B lymphocytes heightened CK2 kinase activity contributes to lymphocyte transformation via its ability to prolong the half-life of c-Myc by reducing degradation of the transcription factor by the proteasome.

The c-Myc transcription factor in *Theileria*-infected B cells mediates its survival role via induction of the anti-apoptotic protein Mcl-1 (Dessauge *et al.* 2005). c-Myc however, can only bind to its specific sites called E-boxes (5′-CACGTG-3′) in the promoters of target genes with the help of a second factor called Max (Dang *et al.* 1999). During Fas-induced apoptosis Max is dephosphorylated and subsequently cleaved by caspase-5 and caspase-7, respectively (Krippner-Heidenreich *et al.* 2001). Thus, in *Theileria*-infected lymphocytes loss of Max due to caspase cleavage would prevent c-Myc from binding to the Mcl-1 promoter aggravating cell

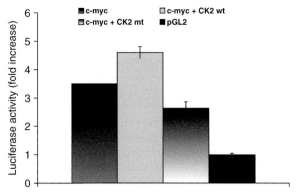

Fig. 2. CK2 over-expression augments c-Myc transactivation in *T. parva*-infected B cells. *T. parva*-infected B cells have constitutive c-Myc activity (c-myc) compared to a control minimal promoter (pGL2) (Dessauge *et al.* 2005). C-Myc-driven luciferase activity is increased upon co-expression with wild type CK2 kinase (c-myc + CK2 wt) and is decreased when co-expressed with a kinase-dead mutant of CK2 (c-myc + CK2 mt). *T. parva*-infected B cells were transfected with a c-myc-luciferase reporter plasmid and constructs expressing wild type and kinase-dead CK2 (Penner *et al.* 1997) and luciferase activity estimated as described (Dessauge *et al.* 2005).

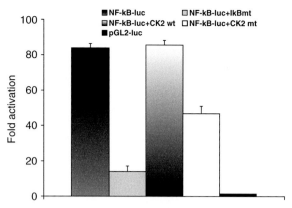

Fig. 3. CK2 contributes to NF-κB activation in *T. parva*-infected B cells. Similar to *T. parva*-infected T cells (Heussler *et al.* 1999), *T. parva*-infected B cells have constitutively high NF-κB-driven transcription (NF-κB-luc). Over-expression of a phosphorylation resistant mutant of IκB (Van Antwerp & Verma, 1996) decreases NF-κB activity (NF-κB-luc + IκBmt) and over-expression of kinase-dead CK2 also reduces NF-κB levels (NF-κB-luc + CK2mt). Over-expression of wild type CK2 kinase does not increase NF-κB (NF-κB-luc + CK2wt) most likely due to all CK2 sensitive sites being phosphorylated. All transfections were performed as described (Dessauge *et al.* 2005).

death. However, cleavage by caspase-5 is inhibited by CK2-mediated phosphorylation of Max (Krippner-Heidenreich *et al.* 2001). So, up-regulation of CK2 in *Theileria*-infected lymphocytes (ole-MoiYoi *et al.* 1993) not only slows down c-Myc degradation, but it could also prevent caspase cleavage of Max and combined these two events would assure sustained transcription of Mcl-1 and lymphocyte survival (Dessauge *et al.* 2005).

CK2 CONTRIBUTES TO NF-κB ACTIVATION IN *THEILERIA*-INFECTED B CELLS

NF-κB is a transcription factor that is activated in response to proinflammatory stimuli, physical stress and in many infections, including those by *Theileria*, where it mediates a survival response (Heussler *et al.* 1999). Activation of NF-κB by many stimuli depends on the IκB kinase (IKK) complex, which phosphorylates IκB (Inhibitor of κB) at N-terminal sites provoking its degradation and in *T. parva*-infected T cells there is continuous phosphorylation and degradation of IκB (Palmer *et al.* 1997). Interestingly, *Theileria* hijacks IKK by sequestering it into a complex associated with the parasite surface (Heussler *et al.* 2002). In UV-induced, NF-κB activation phosphorylation of IκB occurs at a cluster of C-terminal sites that are recognized by CK2 (Kato *et al.* 2003) and CK2 has long been known to phosphorylate IκB (Barroga *et al.* 1995). Thus, it is possible that the high CK2 activity in *T. parva*-infected lymphocytes contributes to IκB phosphorylation, NF-κB activation and lymphocyte survival.

To test this hypothesis, we transiently transfected *T. parva*-infected B cells with wild type and kinase-dead mutants of CK2, together with a luciferase expression construct under the control of NF-κB (Fig. 3). As a control, we independently transfected a phosphorylation resistant mutant of IκB (IBm) to demonstrate that reduced degradation of IκB by the proteasome leads to NF-κB inhibition (Van Antwerp and Verma, 1996). In *T. parva*-infected B cells, over-expression of a kinase-dead mutant of CK2 reduced NF-κB activity by 50%, consistent with a role for CK2 in *Theileria*-dependent NF-κB activation (Fig. 3). Over-expression of active CK2 kinase did not augment the NF-κB activity, most likely due to all the CK2 sensitive sites in IκB being already maximally phosphorylated by the high endogenous CK2 activity of *Theileria*-infected lymphocytes (ole-MoiYoi *et al.* 1993). Thus, CK2 contributes to NF-κB activation by promoting proteasome degradation of IκB and indeed, there could be a positive feedback loop, with activated NF-κB increasing *CK2a* transcription.

The effect of CK2 on c-Myc induction in *Theileria*-infected lymphocytes could also be operating at the transcription level, as NF-κB is known to contribute to *c-myc* transcription in other cell types (Grumont, Strrasser & Gerondakis, 2002). To validate this notion we co-transfected *T. parva*-infected B cells with the *c-myc*-luciferase reporter construct as previous described (Dessauge *et al.* 2005) and IκBm that resists proteasome degradation (Fig. 4). As a control, we again measured the ability of IκBm to inhibit NF-κB. One can see that IκBm induced

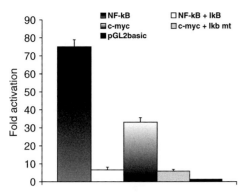

Fig. 4. Over-expression of IκB leads to a significant drop in NF-κB and a partial drop in c-Myc-driven luciferase activity. *T. parva*-infected B cells have constitutively high levels of both NF-κB- (NF-κB) and c-Myc- (c-myc) driven luciferase activities when compared to minimal promoter (pGL2). Over-expression of mutant IκB leads to a significant reduction in NF-κB-driven (NF-κB + IκBmt) and c-Myc-driven (c-myc-luc + IκBmt) luciferase activities demonstrating that NF-κB contributes to transcription of c-myc. All transfections were performed as described (Dessauge *et al.* 2005).

inhibition in NF-κB results in a significant reduction in c-Myc activity. Thus, in *T. parva*-infected B cells NF-κB acts together with STAT3 (Dessauge *et al.* 2005) to induce *c-myc* transcription and CK2 is involved in augmenting c-Myc both at the transcriptional (via NF-κB) and post-translational (via inhibition of proteasome degradation) levels.

CK2 AND THE LIPID PHOSPHATASE PTEN

The PTEN tumour suppressor gene encodes a phosphatidylinositol 3′-phosphatase that is inactivated in a high percentage of human tumours and its inactivation leads to constitutive phosphatidylinositol 3′-kinase (PI3-K) activity. *Theileria*-infected lymphocytes are characterized by high constitutive levels of PI3-K (Baumgartner *et al.* 2000; Heussler *et al.* 2001) suggestive of inactivation of PTEN. However, upon drug-induced parasite death PI3-K activity drops implying that PTEN is not permanently inactivated by mutation. Previous studies in other cellular systems have shown that Ser(370), Ser(385) and Thr(366) of PTEN are phosphorylated by CK2 and their phosphorylation inhibits PTEN activity towards PIP3 produced by PI3-K (Miller *et al.* 2002).

PTEN can also be inactivated by caspase-3 cleavage at several sites located within the C terminus of the molecule and this increased when cells are stimulated with TNF-α, but caspase-3 cleavage is negatively regulated by phosphorylation of the C-terminal tail of PTEN by the protein kinase CK2 (Torres *et al.* 2003). This probably occurs in *T. parva*-infected B cells, as PTEN appears intact (see Fig. 5), even though they use a TNF-autocrine

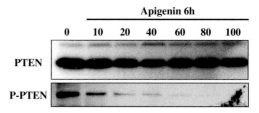

Fig. 5. Pharmacological inhibition of CK2 by Apigenin leads to loss of PTEN phosphorylation. Infected B cells were treated for 6 hours with increasing doses of Apigenin (from 0 to 100 μM), as described in (Dessauge *et al.* 2005). Inhibition of CK2 ablates PTEN phosphorylation without effecting PTEN levels.

loop for proliferation (Guergnon *et al.* 2003*a*). The high CK2 activity of *Theileria*-infected lymphocytes therefore, appears to counterbalance any TNF-induced, caspase-3-mediated cleavage of PTEN. Thus, CK2 has both positive (phosphorylation blocking caspase 3 cleavage) and negative (phosphorylation inhibiting PTEN phosphates activity) effects on PTEN, the overall balance in *Theileria*-infected cells leading to a loss of phosphatase activity that results in the permanent PI3-K induction that characterizes both infected B and T cells (Baumgartner *et al.* 2000; Heussler *et al.* 2001).

Interestingly, Pten heterozygous (Pten+/−) mutant mice that have reduced phosphatase activity develop a lethal polyclonal autoimmune disorder (homozygous Pten −/− mice are embryonic lethal) with features reminiscent of those observed in Fas-deficient mice (Di Cristofano *et al.* 1999). Fas-mediated apoptosis is impaired in Pten +/− mice and T cells show reduced activation-induced cell death and increased proliferation upon activation and inhibition of PI3-K restored Fas responsiveness in Pten +/− cells. In this context, it is worth pointing out that *T. parva*-infected T cells also show resistance to Fas-mediated apoptosis (Kuenzi, Schneider & Dobbelaene, 2003) and this observation is entirely consistent with the notion of CK2-mediated inhibition of PTEN.

CK2 AND REGULATION OF CASPASE ACTIVITY

As stated, *T. parva*-infected B cells use a TNF autocrine loop to proliferate and inhibition of TNF did not generate apoptosis (Guergnon *et al.* 2003*a*). Usually TNF-related apoptosis-inducing ligand (TRAIL) induces apoptosis via the death receptors DR4 and DR5 in transformed cells and inhibition of CK2 results in sensitization of tumour cells to TRAIL-induced apoptosis (Hilgard *et al.* 2004). Analysis of the death-inducing signaling complex (DISC) demonstrated that CK2 activation diminishes recruitment of procaspase-8 to the DISC and decreases degradation of X-linked inhibitor of apoptosis protein (XIAP). In *T. parva*-infected T cells XIAP is constitutively expressed and upon drug-induced parasite death XIAP levels drop (Kuenzi *et al.* 2003). This could be interpreted as being due to loss of CK2 activity that also decreases upon parasite death (ole-MoiYoi *et al.* 1993). In addition, resistance to TRAIL-mediated apoptosis in tumour cells is associated with a high Bcl-x(L)/tBID ratio and failure to activate caspase-9 (Ravi & Bedi, 2002). In *T. parva*-infected B cells and CD8+ T cells drug-induced parasite death leads to induction of caspase 9 and an increase in pro-apoptotic Bax, compared to anti-apoptotic Bcl-2 (Guergnon *et al.* 2003*b*). So, to determine whether, in infected T cells or B cells, parasite-dependent CK2-mediated phosphorylation appears to modulate caspase activity and to gain support for this idea we measured the activity of caspase 3 in *T. parva*-infected B cells that had been treated with increasing doses of two different CK2 inhibitors (Fig. 6). One can observe that even in the presence of live parasites inhibition of CK2 leads to induction of caspase 3. So, CK2 plays an anti-apoptotic role not only via induction of NF-κB and c-Myc, but also via suppression of caspase activation.

CK2 AND THE SERINE/THREONINE PHOSPHATASE PP2A

The dynamic process of signal transduction involves the concerted action of both protein kinases and protein phosphatases. Among serine/threonine phosphatases, PP1 and PP2A are key regulators of protein phosphorylation being responsible for more than 90% of protein dephosphorylation. PP2A is the predominant phosphatase activity detected in *T. parva*-infected B cells, whereas PP1 is specifically associated with the parasite (Cayla *et al.* 2000). Interestingly, PP2A can be activated by interaction with the alpha subunit of CK2 (Heriche *et al.* 1997) and activation of the phosphatase resulted in de-activation of mitogen-activated protein kinase kinase (MEK). Thus, CK2A acts as a stimulatory PP2A subunit, in contrast to the small t antigen of SV40 virus that behaves as an inhibitory subunit of PP2A (Sontag *et al.* 1993). In direct contrast to PP2A stimulation by CK2A, inhibition of the phosphatase by SV40 small t antigen leads to constitutive activation of the MEK/ERK pathways in virus-transformed cells (Sontag *et al.* 1993). In *Theileria*-infected lymphocytes the MEK/ERK pathway is constitutively switched off, consistent with CK2A stimulation of PP2A (Chaussepied & Langsley, 1996; Galley *et al.* 1997; Chaussepied *et al.* 1998). Therefore, it is attractive to imagine that in *Theileria*-infected lymphocytes there is an association between CK2A and PP2A that activates the phosphatase and leads to permanent dephosphorylation of MEK/ERK.

The MEK/ERK pathway may not be the only one down-regulated by a putative association between

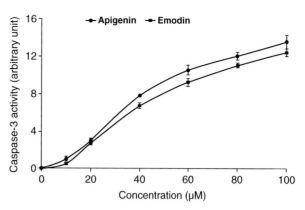

Fig. 6. Activation of caspase 3 increases following inhibition of CK2. Infected B cells were treated for 6 hours with increasing doses of Apigenin and Emodin (from 0 to 100 μM) and caspase-3 activity generated following cleavage of pro-caspase 3 was assayed by measuring cleavage of DEVD-AMC, as described in (Dessauge *et al.* 2005).

CK2A and PP2A, as P22A has also been described to function as a ser473 phosphatase of Akt/PKB (Liu *et al.* 2003) and lack of ser473 phosphorylation of Akt is observed in *T. parva*-infected B cells (Baumgartner *et al.* 2000). Therefore, the putative PP2A/CK2A phosphatase might be responsible for both lack of ERK activity and poor Akt/PKB activity associated by *T. parva*-infected B cells.

CONCLUSIONS

Activation of CK2 was one of the first observations made on how *Theileria* parasites manipulate host cell signal transduction pathways (ole-MoiYoi *et al.* 1993). In this review, we argue that CK2 induction may in fact contribute to many of the different activation events that have been described since 1993 for *Theileria*-infected lymphocytes and as such this makes CK2 a central player in lymphocyte transformation, a position that we have resumed graphically in Fig. 7. Most of our examples concern phosphorylation by CK2 and it can lead to sustained activation of transcription factors c-Myc (Dessauge *et al.* 2005) and NF-κB (Heussler *et al.* 1999), but by slightly different mechanisms. In the case of c-Myc, CK2 phosphorylation of a PEST sequence at the C-terminus reduces proteasome-mediated degradation of the transcription factor (Channavajhala & Seldin, 2002) and results in more robust transactivation of c-Myc-mediated transcription (Fig. 2). In the case of NF-κB, it is the CK2 phosphorylation of I-κB (Barroga *et al.* 1995) that promotes proteasome degradation of the inhibitor of NF-κB and enhanced NF-κB-mediated transcription (Fig. 3). Moreover, since diminished degradation of IκB also contributes to c-Myc induction via NF-κB activation (Fig. 4), CK2 increases c-Myc induction both directly and indirectly. As both c-Myc and NF-κB mediate

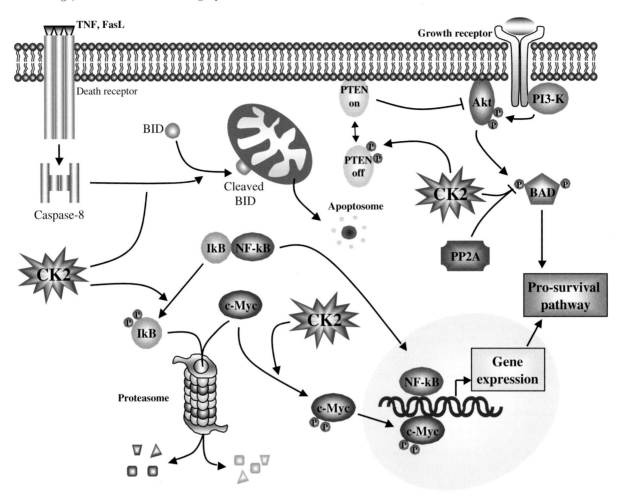

Fig. 7. Scheme showing central role that CK2 could play in *Theileria*-induced lymphocyte transformation. CK2 could dampen TNF/Fas death signalling by blocking caspase 8 cleavage of Bid and the release of the apoptosome from mitochondria. CK2 phosphorylation accelerates IκB degradation by the proteasome and hence augments NF-κB activation. CK2 phosphorylation ablates c-Myc degradation by the proteasome. CK2 phosphorylation of PTEN inhibits its lipid phosphatase activity and maintains PI3-K activity at high levels. CK2 could interact with PP2A and stimulate the protein phosphatase to dephosphorylate BAD and Akt. A combination of increased transcriptional transaction of NF-κB and c-Myc, combined with inhibition of apoptosis allows pro-survival signalling to dominate in *Theileria*-infected lymphocytes.

anti-apoptotic responses of *Theileria*-infected lymphocytes, CK2 activation contributes to survival of infected host cells.

CK2 also contributes to infected lymphocyte survival by inhibiting caspase activation (Fig. 6) and this probably explains the resistance of *T. parva*-infected T cells to Fas-mediated apoptosis (Kuenzi *et al.* 2003). CK2 activation diminishes both recruitment of procaspase-8 to the DISC and decreases degradation of survival protein XIAP (Hilgard *et al.* 2004). CK2 activation could also explain why inhibition of TNF does not lead to apoptosis of *T. parva*-infected B cells (Guergnon *et al.* 2003a), via its role in dampening TRAIL/Apo2-mediated apoptosis (Ravi & Bedi, 2002; Farah *et al.* 2003).

CK2 is probably behind constitutive PI3-K activation that although this does not mediate an anti-apoptotic response, since it does not signal to NF-κB (Heussler *et al.* 2001), it does contribute to

proliferation and AP-1 induction (Baumgartner *et al.* 2000). CK2 does this by phosphorylating PTEN (Fig. 5) an event that leads to a reduction in its lipid phosphatase activity and higher PIP3 levels, where PIP3 is the product of PI3-K (Miller *et al.* 2002). Permanently high PIP3 levels could explain how PI3-K contributes to proliferation, AP-1 and GM-CSF induction in *T. parva*-infected B cells (Baumgartner *et al.* 2000).

Finally, we give an example of where CK2A may be acting not only as a kinase, but also as a stimulatory subunit for the protein phosphatase PP2A (Heriche *et al.* 1997). In *T. parva*-infected B cells PP2A is the major protein phosphatase activity (Cayla *et al.* 2000) and CK2A activated PP2A could explain the lack MEK/ERK (Galley *et al.* 1997; Chaussepied *et al.* 1998) and reduced Akt/PKB activities associated with infected B cells (Baumgartner *et al.* 2000). Dampening down the MEK/ERK and

Akt/PKB pathways may be as essential to successful lymphocyte survival as activating the c-Myc and NF-κB pathways and that is why we propose CK2 as a central player in *Theileria*-induced lymphocytes transformation.

ACKNOWLEDGMENTS

We thank Volker Heussler for critical comment and David Litchfield for the gift of the wild type and mutants CK2 kinase constructs and Inder Verma for the phosphorylation resistant mutant of IκB. We acknowledge the support of the CNRS and INSERM.

REFERENCES

BARROGA, C. F., STEVENSON, J. F., SCHWARZ, E. M. & VERMA, I. M. (1995). Constitutive phosphorylation of I kappa B alpha by casein kinase II. *Proceedings of the National Academy of Sciences, USA* **92**, 7637–7641.

BAUMGARTNER, M., CHAUSSEPIED, M., MOREAU, M. F., WERLING, D., DAVIS, W. C., GARCIA, A. & LANGSLEY, G. (2000). Constitutive P13-K activity is essential for proliferation, but not survival, of *Theileria parva*-transformed B cells. *Cellular Microbiology* **2**, 329–339.

BIERMANN, R., SCHNITTGER, L., BEYER, D. & AHMED, J. S. (2003). Initiation of translation and cellular localization of *Theileria annulata* casein kinase IIalpha: implication for its role in host cell transformation. *Journal of Cell Physiology* **196**, 444–453.

CAYLA, X., GARCIA, A., BAUMGARTNER, M., OZON, R. & LANGSLEY, G. (2000). A *Theileria parva* type 1 protein phosphatase activity. *Molecular and Biochemical Parasitology* **110**, 161–166.

CHANNAVAJHALA, P. & SELDIN, D. C. (2002). Functional interaction of protein kinase CK2 and c-Myc in lymphomagenesis. *Oncogene* **21**, 5280–5288.

CHAUSSEPIED, M., LALLEMAND, D., MOREAU, M. F., ADAMSON, R., HALL, R. & LANGSLEY, G. (1998). Upregulation of Jun and Fos family members and permanent JNK activity lead to constitutive AP-1 activation in *Theileria*-transformed leukocytes. *Molecular and Biochemical Parasitology* **94**, 215–226.

CHAUSSEPIED, M. & LANGSLEY, G. (1996). *Theileria* transformation of bovine leukocytes: a parasite model for the study of lymphoproliferation. *Research in Immunology* **147**, 127–138.

DANG, C. V., RESAR, L. M., EMISON, E., KIM, S., LI, Q., PRESCOTT, J. E., WONSEY, D. & ZELLER, F. (1999). Function of the c-Myc oncogenic transcription factor. *Experimental Cell Research* **253**, 63–77.

DESSAUGE, F., HILALY, S., BAUMGARTNER, M., BLUMEN, B., WERLING, D. & LANGSLEY, G. (2005). c-Myc activation by *Theileria* parasites promotes survival of infected B-lymphocytes. *Oncogene*, **24**, 1075–1083.

DI CRISTOFANO, A., KOTSI, P., PENG, Y. F., CORDON-CARDO, C., ELKON, K. B. & PANDOLFI, P. P. (1999). Impaired Fas response and autoimmunity in Pten +/− mice. *Science* **285**, 2122–2125.

FARAH, M., PARHAR, F., MOUSSAVI, M., EIVEMARK, S. & SALH, B. (2003). 5,6-Dichloro-ribifuranosylbenzimidazole- and apigenin-induced sensitization of colon cancer cells to TNF-alpha-mediated apoptosis. *American Journal*

of Physiology Gastrointest Liver Physiology **285**, G919–928.

FILHOL, O., MARTIEL, J. L. & COCHET, C. (2004). Protein kinase CK2: a new view of an old moolecular complex. *EMBO Rep* **5**, 351–355.

GALLEY, Y., HAGENS, G., GLASER, I., DAVIS, W., EICHHORN, M. & DOBBELAERE, D. (1997). Jun NH2-terminal kinase is constitutively activated in T cells transformed by the intracellular parasite *Theileria parva*. *Proceedings of the National Academy of Sciences, USA* **94**, 5119–5124.

GRUMONT, R. J., STRASSER, A. & GERONDAKIS, S. (2002). B cell growth is controlled by phosphatidylinosotol 3-kinase-dependent induction of Rel/NF-kappaB regulated c-myc transcription. *Molecular Cell* **10**, 1283–1294.

GUERGNON, J., CHAUSSEPIED, M., SOPP, P., LIZUNDIA, R., MOREAU, M. F., BLUMEN, B., WERLING, D., HOWARD, C. J. & LANGSLEY, G. (2003*a*). A tumour necrosis factor alpha autocrine loop contributes to proliferation and nuclear factor-kappaB activation of *Theileria parva*-transformed B cells. *Cellular Microbiology* **5**, 709–716.

GUERGNON, J., DESSAUGE, F., LANGSLEY, G. & GARCIA, A. (2003*b*). Apoptosis of *Theileria*-infected lymphocytes induced upon parasite death involves activation of caspases 9 and 3. *Biochimie* **85**, 771–776.

HERICHE, J. F., LEBRIN, F., RABILLOUD, T., LEROY, D., CHAMBAZ, E. M. & GOLDBERG, Y. (1997). Regulation of protein phosphatase 2A by direct interaction with casein kinase salpha. *Science* **276**, 952–955.

HEUSSLER, V. T., KUENZI, P., FRAGA, F., SCHWAB, R. A., HEMMINGS, B. A. & DOBBELAERE, D. A. (2001). The Akt/PKB pathway is constitutively activated in *Theileria*-transformed leucocytes, but does not directly control constitutive NF-kappaB activation. *Cellular Microbiology* **3**, 537–550.

HEUSSLER, V. T., MACHADO, J., jr., FERNANDEZ, P. C., BOTTERON, C., CHEN, C. G., PEARSE, M. J. & DOBBELAERE, D. A. (1999). The intracellular parasite *Theileria parva* protects infected T cells from apoptosis. *Proceedings of the National Academy Sciences, USA* **96**, 7312–7317.

HEUSSLER, V. T., ROTTENBERG, S., SCHWAB, R., KUENZI, P., FERNANDEZ, P. C., McKELLAR, S., SHIELS, B., CHEN, Z. J., ORTH, F., WALLACH, D. & DOBBELAERE, D. A. (2002). Hijacking of host cell IKK signalosomes by the transforming parasite *Theileria*. *Science* **298**, 1033–1036.

HILGARD, P., CZAJA, M. J., GERKEN, G. & STOCKERT, R. J. (2004). Proapoptotic function of protein kinase CK2alpha is mediated by a JNK signaling cascade. *American Journal of Physiol Gastrointest Liver Physiol* **287**, G192–201.

KATO, T., Jr., DELHASE, M., HOFFMANN, A. & KARIN, M. (2003). CK2 Is a C-terminal IkappaB kinase responsible for NF-kappaB activation during the UV response. *Mol Cell*, **12**, 829–839.

KREHAN, A., ANSUINI, H., BOCHER, O., GREIN, S., WIRKNER, U. & PYERIN, W. (2000). Transcription factors ets1, NF-kappa B, and Sp1 are major determinants of the promoter activity of the human protein kinase CK2alpha gene. *Journal of Biological Chemistry* **275**, 18327–18336.

KRIPPNER-HEIDENREICH, A., TALANIAN, R. V., SEKUL, R., KRAFT, R., THOLE, H., OTTLEBEN, H. & LUSCHER, B. (2001). Targeting of the transcription factor Max during apoptosis: phosphorylation-regulated cleavage by caspase-5 at an unusual glutamic acid residue in position P1. *Biochemical Journal* **358**, 705–715.

KUENZI, P., SCHNEIDER, P. & DOBBELAERE, D. A. (2003). *Theileria parva*-transformed T cells show enhanced resistance to Fas/Fas ligand-induced apoptosis. *Journal of Immunology* **171**, 1224–1231.

LIU, W., AKHAND, A. A., TAKEDA, F., KAWAMOTO, Y., ITOIGAWA, M., KATO, M., SUZUKI, H., ISHIKAWA, N. & NAKASHIMA, I. (2003). Protein phosphatase 2A-linked and-unlinked caspase-dependent pathways for downregulation of Akt kinase triggered by 4-hydroxynonenal. *Cell Death and Differentiation* **10**, 772–781.

MILLER, S. J., LOU, D. Y., SELDIN, D. C., LANE, W. S. & NEEL, B. G. (2002). Direct identification of PTEN phosphorylation sites. *FEBS Letters* **528**, 145–53.

NIEFIND, F., GUERRA, B., ERMAKOWA, I. & ISSINGER, O. G. (2001). Crystal structure of human protein kinase CK2: insights into basic properties of the CK2 holoenzyme. *EMBO Journal* **20**, 5320–5331.

OLE-MOIYOI, O. F. (1995). Casein kinase II in theileriosis. *Science* **267**, 834–836.

OLE-MOIYOI, O. F., BROWN, W. C., IAMS, K. P., NAYAR, A., TSUKAMOTO, T. & MACKLIN, M. D. (1993). *Theileria parva*: an intracellular protozoan parasite that induces reversible lymphocyte transformation. *EMBO Journal* **12**, 1621–1631.

OLE-MOIYOI, O. F., SUGIMOTO, C., CONRAD, P. A. & MACKLIN, M. D. (1992). Cloning and characterization of the casein kinase II alpha subunit gene from the lymphocyte-transforming intracellular protozoan parasite. *Theileria parva*. *Biochemistry* **31**, 6193–6202.

PALMER, G. H., MACHADO, J., Jr., FERNANDEZ, P., HEUSSLER, V., PERINAT, T. & DOBBELAERE, D. A. (1997). Parasite-mediated nuclear factor kappaB regulation in lymphoproliferation caused by *Theileria parva* infection.

Proceedings of the National Academy Sciences, USA **94**, 12527–12532.

PENNER, C. G., WANG, Z. & LITCHFIELD, D. W. (1997). Expression and localization of epitope-tagged protein kinase CK2. *Journal of Cellular Biochemistry* **64**, 525–537.

RAVI, R. & BEDI, A. (2002). Sensitization of tumor cells to Apo2 ligand/TRAIL-induced apoptosis by inhibition of casein kinase II. *Cancer Research* **62**, 4180–4185.

SAYED, M., KIM, S. O., SALH, B. S., ISSINGER, O. G. & PELECH, S. L. (2000). Stress-induced activation of protein kinase CK2 by direct interaction with p38 mitogen-activated protein kinase. *Journal of Biological Chemistry* **275**, 16569–16573.

SELDIN, D. C. & LEDER, P. (1995). Casein kinase II alpha transgene-induced murine lymphoma: relation to theileriosis in cattle. *Science* **267**, 894–897.

SHAYAN, P. & AHMED, J. S. (1997). *Theileria*-mediated constitutive expression of the casein kinase II-alpha subunit in bovine lymphoblastroid cells. *Parasitology Research* **83**, 526–532.

SONTAG, E., FEDOROV, S., KAMIBAYASHI, C., ROBBINS, D., COBB, M. & MUMBY, M. (1993). The interaction of SV40 small tumor antigen with protein phosphatase 2A stimulates the map kinase pathway and induces cell proliferation. *Cell* **75**, 887–897.

TORRES, J., RODRIGUEZ, J., MYERS, M. P., VALIENTE, M., GRAVES, J. D., TONKS, N. F. & PULIDO, R. (2003). Phosphorylation-regulated cleavage of the tumor suppressor PTEN by caspase-3: implications for the control of protein stability and PTEN-protein interactions. *Journal of Biological Chemistry* **278**, 30652–30660.

VAN ANTWERP, D. J. & VERMA, I. M. (1996). Signal-induced degradation of I(kappa)B(alpha): association with NF-kappaB and the PEST sequence in I(kappa)B(alpha) are not required. *Molecular Cell Biology* **16**, 6037–6045.

Stimulation of innate immune responses by malarial glycosylphosphatidylinositol via pattern recognition receptors

T. NEBL, M. J. DE VEER *and* L. SCHOFIELD*

The Walter and Eliza Hall Institute of Medical Research, 1G Royal Parade, Victoria, 3050, Australia

SUMMARY

The glycosylphosphatidylinositol (GPI) anchor of *Plasmodium falciparum* is thought to function as a critical toxin that contributes to severe malarial pathogenesis by eliciting the production of proinflammatory responses by the innate immune system of mammalian hosts. Analysis of the fine structure of *P. falciparum* GPI suggests a requirement for the presence of both core glycan and lipid moieties in the recognition and signalling of parasite glycolipids by host immune cells. It has been demonstrated that GPI anchors of various parasitic protozoa can mediate cellular immune responses via members of the Toll-like family of pattern recognition receptors (TLRs). Recent studies indicate that GPI anchors of *P. falciparum* and other protozoa are preferentially recognized by TLR-2, involving the MyD88-dependent activation of specific signalling pathways that mediate the production of proinflammatory cytokines and nitric oxide from host macrophages *in vitro*. However, the contribution of malaria GPI toxin to severe disease syndromes and the role of specific TLRs or other pattern recognition receptors in innate immunity *in vivo* is only just beginning to be characterized. A better understanding of the molecular mechanisms underlying severe malarial pathogenesis may yet lead to substantial new insights with important implications for the development of novel therapeutics for malaria treatment.

Key words: Malaria, GPI, *P. falciparum*, *T. cruzi*, *T. brucei*, *T. gondii*, pathology, Toll receptors, pattern recognition.

INTRODUCTION: MALARIA DISEASE

Malaria is an infectious disease that causes enormous global medical and economic burdens. Of the four species of malaria parasites known to infect humans, the bulk of severe disease and complications is caused by the species *Plasmodium falciparum*. More than 80% of the approximately 2 million estimated cases of malaria-related mortality worldwide occur in Africa, mostly affecting infants and pre-school children aged 6 months to 5 years. Malarial fatalities are associated with a spectrum of discrete and overlapping disease syndromes of complex aetiologies. Humans affected and dying of malaria may variously suffer systemic, single- or multi-organ involvement, including acute respiratory distress, coagulopathy, shock, metabolic acidosis, hypoglycaemia, renal failure, pulmonary oedema and cerebral involvement including seizures and coma (White & Ho, 1992). Basic mechanisms controlling these processes are thought to be the site-specific localization of parasites (by cytoadherence to vascular endothelial markers such as the adhesin ICAM-1 (Berendt *et al.* 1989), and both local and systemic inflammatory responses arising from the action of cytokines produced predominantly by the

innate immune system (Stevenson & Riley, 2004). The ability of the intraerythrocytic parasites to cytoadhere to endothelium and sequester in the microvasculature of vital organs (Miller *et al.* 2002) is thought to promote the accumulation of high local concentrations of bioactive parasite products (GPI-anchored antigens and glycolipids, metabolite toxins) that can trigger a cascade of pro-inflammatory immune responses in the brain, lungs, spleen, liver, kidney and placenta. This cytokine-dependent inflammatory cascade is believed to contribute to the spectrum of acute clinical symptoms that manifest during severe malaria bouts. In order to better understand the molecular mechanisms underlying severe malarial pathogenesis, our group has focused its attention on the interaction between *P. falciparum* and the immune system of the vertebrate host in the context of malaria infection.

MOLECULAR BASIS OF SEVERE MALARIA – CYTOKINE CONTRIBUTIONS

When examining the molecular basis of malarial pathogenesis it is clear that many factors from both the parasite and the host immune system contribute to the development of severe disease and there is an intricate relationship between these factors (reviewed by Chen, Schlichtherle & Wahlgren 2000; Stevenson & Riley, 2004). Both human and murine malaria infections elicit cells of the innate immune

* Corresponding author: Louis Schofield, The Walter and Eliza Hall Institute of Medical Research, 1G Royal Parade, Victoria 3050, Australia. Phone: (613) 9345 2474 Fax: (613) 9347 0852. E-mail:schofield@wehi.edu.au

Parasitology (2005), **130**, S45–S62. © 2005 Cambridge University Press
doi:10.1017/S0031182005008152 Printed in the United Kingdom

compartment to produce high levels of pro-inflammatory cytokines such as interferon-gamma (IFN-γ), interleukin-1 (IL-1), IL-6, IL-12 and tumor necrosis factor alpha (TNF-α) (Grau *et al.* 1987, 1989*b*; Lucas *et al.* 1997; Lyke *et al.* 2004). These pro-inflammatory cytokines appear to be essential for mediating acquired immunity to the invasive sporozoite during the early stages of infection (Schofield *et al.* 1987), but their role in mediating host resistance to blood stage infection is less clear.

The over-production of inflammatory cytokines can also be harmful to the host. Serum IFN-γ levels are elevated in patients with acute *P. falciparum* malaria (Ringwald *et al.* 1991), and death resulting from cerebral malaria (CM) was found to be associated with heterozygosity for an IFN-γ receptor-1 polymorphism in Gambian children (Koch *et al.* 2002). In the murine model both IFN-γ and IFN-γ receptor knock-out mice are resistant to *Plasmodium berghei*-mediated CM (Amani *et al.* 2000), and injection of neutralizing monoclonal antibody (mAb) against IFN-γ was shown to protect against CM in C57BL/6 susceptible mice (Grau *et al.* 1989*a*). These data suggest a role for IFN-γ as a key regulator of downstream immune system processes in the pathogenesis of severe malaria. IFN-γ has multiple activities in both the innate and acquired compartments of the immune system. Importantly, IFN-γ induces macrophages to become highly activated and, when exposed to live parasites, to generate high levels of free radical nitrogen intermediates (RNIs) via an inducible nitric oxide synthase (iNOS). RNIs can cause irreversible damage to neuronal tissues and have been implicated in the pathogenesis of CM (Al Yaman *et al.* 1996). In addition, when stimulated with parasite products *in vitro*, IFN-γ-primed macrophages can produce high levels of other pro-inflammatory cytokines associated with severe malaria, including IL-1, IL-6, IL-10, IL-12 and TNF-α (Grau *et al.* 1987, 1989*b*; Lucas *et al.* 1997; Lyke *et al.* 2004). Severe malaria is also associated with polymorphisms in the TNF-α promoter region, which may be linked to increased TNF-α expression (Mcguire *et al.* 1994), and high circulating levels of TNF-α are a prognostic indicator of fatality (Grau *et al.* 1989*b*). In concert, IFN-γ and TNF-α cytokines are capable of synergistically up-regulating intracellular adhesion molecules such as ICAM-1, VCAM-1, CD36 and thrombospondin on various cell types, including vascular endothelium and leukocytes. This augments sequestration of parasitized erythrocytes to endothelium and intensifies the deposition of malaria toxins in the deep vasculature of organs such as the brain. Malaria fatalities are thus strongly associated with an exacerbated systemic or organ-specific inflammatory cascade, but the molecular nature of the parasite toxin(s) that trigger the overproduction of pro-inflammatory cytokines has for a long time remained unclear.

MALARIAL GPIS AS CANDIDATE TOXINS AND PATHOGENICITY FACTORS

Several years ago, *in vitro* and *in vivo* studies suggested that glycosylphosphatidylinositol (GPI)-anchored molecules of *P. falciparum* function as the dominant malarial toxin in the context of infection. This hypothesis was based on the finding that purified GPI-anchored glycoproteins and glycolipids of *P. falciparum* and *P. berghei* are very potent activators of innate immune cells, capable of inducing pro-inflammatory cytokines TNF-α and IL-1 from macrophages (Schofield & Hackett, 1993). This activity has subsequently been confirmed by others (Naik *et al.* 2000; Vijaykumar, Naik & Gowda, 2001) (Table 1). The original study also showed that when administered to mice, GPI alone is sufficient to cause symptoms similar to acute malaria infection such as transient fever and hypoglycemia and death of recipients due to TNF-α-mediated sepsis (Schofield *et al.* 1993). Initial data showed that the endotoxin activity of major *P. falciparum* surface antigens released during schizont rupture (i.e. MSP-1, MSP-2) is restricted to the GPI moiety, since protein denaturation and exhaustive digestion with pronase does not abolish or reduce the potency of these molecules for the induction of cytokines *in vitro* (Schofield & Hackett, 1993). In contrast, enzymatic cleavage with phospholipase A_2 or chemical treatment of the GPI moiety with mild alkali (for selective removal of fatty acid esterified at the glycerol *sn*-1 and *sn*-2 postitions) or nitrous acid (for selective breakage of the glucosamine-*myo*-inositol bond) reduces most of the pro-inflammatory activity (Schofield & Hackett, 1993) – indicating a requirement for intact GPI-associated glycolipid structure. Additional studies found that purified *P. falciparum* GPI glycolipids were also able to induce, either directly or in synergy with IFN-γ, a range of other activities in various host tissues and cell types, including insulin-mimetic signalling in adipocytes (Schofield *et al.* 1994, 1996), the induction of iNOS in macrophages and vascular endothelium (Tachado *et al.* 1996), and the up-regulation of cell adhesion molecules on the surface of host leukocytes and endothelial cells (Schofield *et al.* 1996). The bioactivity of purified GPI glycolipid was comparable to that obtained with live parasites *in vitro*, in the range of 0·1–10 parasite equivalents per macrophage (Schofield & Hackett, 1993; Schofield *et al.* 1996). Importantly, monoclonal or polyclonal antibodies that recognize malarial GPI completely block the induction of TNF-α or iNOS output from macrophages and ICAM-1 expression on endothelial cells by crude extracts of *P. falciparum* (Schofield *et al.* 1994, 1996, 2002). Thus, *Plasmodium* GPIs appear both necessary and sufficient for the induction of pro-inflammatory host responses by malarial parasites *in vitro*.

Table 1. Support for pro-inflammatory bioactivity of protozoal GPIs

GPI preparation	Activity	Reference
GPI purified from pronase-digested *P. falciparum* MSP-1 and MSP-2 and *T. brucei* VSG	Induction of TNF-α and IL-1 production in mouse peritoneal macrophages	Schofield & Hackett (1993) Tachado & Schofield (1994)
Octylsepharose and TLC-purified GPIs from *P. falciparum*	Induction of iNOS and ICAM expression in mouse peritoneal macrophages and human vascular endothelium cells	Tachado *et al.* (1996) Schofield *et al.* (1996)
Octylsepharose and TLC-purified *P. falciparum* and *T. brucei* GPIs	Activation of PTK, PKC and NF-κB signalling pathways and TNF-α, IL-1 and iNOS production in murine RAW264 macrophages	Tachado *et al.* (1997)
Octylsepharose-purified *T. cruzi* trypomastigote mucin GPIs	TNF-α and iNOS output in IFN-γ-primed peritoneal macrophages from C3H/HeJ mice	Camago *et al.* (1997*a,b*)
Highly purified GPIs and GIPLs of *T. cruzi* trypomastigote forms	TNF-α, IL-12 and iNOS production in peritoneal macrophages from LPS-hyposensitive C3H/HeJ mice	Almeida *et al.* (2000)
HPLC and TLC-purified *P. falciparum* GPIs	TNF-α output in murine J774A.1 macrophages	Vijaykumar *et al.* (2001) Naik *et al.* (2000)
Purified *T. brucei* VSG GPI anchors	TNF-α and IL-1 production in IFN-γ-primed peritoneal macrophages from C3H/HeJ mice and bovine peripheral blood monocytes	Magez *et al.* (1998) Sileghem *et al.* (2001)
HPTLC-purified and chemically synthesized GPIs of *Toxoplasma gondii*	TNF-α production in murine RAW264 macrophages	Debierre-Grockiego *et al.* (2003)
Octylsepharose-purified *T. cruzi* mucin GPI anchors	Activation of TLR-2, MAPK and NF-κB signalling pathways required for IL-12, TNF-α and iNOS production in mouse macrophages and gene induction in transfected CHO cells	Ropert *et al.* (2001) Campos *et al.* (2001)
HPLC and HPTLC-purified *P. falciparum* GPIs	Activation of TLR-2/MyD88, MAPK, JNK and NF-κB signalling pathways required for the expression of TNF-α, IL-6, IL-12 and iNOS in murine bone marrow-derived macrophages and human peripheral blood monocytes	Krishnegowda *et al.* (2005) Zhu *et al* .(2005)

These findings were then extended to GPIs of other parasitic protozoa (Table 1). For example, the GPI anchor of *Trypanosoma brucei* VSG and highly purified GPIs of *Trypanosoma cruzi* trypomastigote origin are potent macrophage activators (Tachado & Schofield, 1994; Magez *et al.* 1998; Almeida & Gazzinelli, 2001), and this may account for many features of pathogenesis of trypanosomiasis and Chagas disease. More recently, it was shown that highly purified GPIs of *Toxoplasma gondii* tachyzoites, the infective form of the parasite which causes severe encephalitis, are bioactive factors that participate in the production of inflammatory cytokines from macrophages during toxoplasmal pathogenesis (Debierre-Grockiego *et al.* 2003). In contrast, GIPLs and lipophosphoglycans derived from infective promastigote forms of *Leishmania* spp. suppress several functions of the host immune system – thereby allowing this parasite to evade the innate

immunological control of infection (Camargo *et al.* 1997*b*; Tachado *et al.* 1997). Taken together, these data support the view that GPIs of the parasitic protozoa are both immunostimulatory and immunoregulatory components in the context of protozoal infections and the important differences in bioactivity that exist depend upon GPI fine structure.

GPI ACTS AS A CRITICAL TOXIN IN THE CONTEXT OF SEVERE DISEASE SYNDROMES

The data reviewed so far indicate that malaria GPI is sufficient to act as a toxin but do not establish that it plays this role in the context of authentic disease processes. We therefore sought to test this proposition in a credible pre-clinical model of disease. *P. berghei* ANKA murine malaria has salient features in common with several aspects of the human severe and cerebral malaria syndromes and is thus an accepted model for certain important aspects of the human disease (De Souza & Riley, 2002). It manifests a cytokine-dependent encephalopathy associated with up-regulation of adhesins on the cerebral microvascular endothelium and attendant neurological complications (Grau *et al.* 1987, 1989*a*; 1991; Jennings *et al.* 1997). The more generalized syndromes of pulmonary oedema, lactic acidosis, coagulopathies, shock and renal failure are also observed associated with fatalities (Chang *et al.* 2001), indicating that the *P. berghei* ANKA infection models many aspects of clinically severe malaria beyond narrow definitions of cerebral symptomology. Unlike most, but not all, human cerebral malaria cases, however, there is a macrophage infiltrate and compromised blood-brain barrier in the terminal or agonal stages of the murine syndrome. Nonetheless, in the proximal or developmental stages the murine disease is accepted to reflect more accurately the cytokine-dependent inflammatory cascade leading to cerebral and systemic involvement in humans. As recently reviewed (De Souza & Riley, 2002; Miller *et al.* 2002), *P. berghei* ANKA thus appears the best available small animal model of clinically severe malaria, suitable for investigations into proximal events controlling the systemic inflammatory cascade, and particularly the regulation of cytokine-dependent events and pathology.

Chemical synthesis of the non-toxic glycan moiety of mature *Plasmodium* GPI has recently provided a means to test the hypothesis that malaria GPI is causally involved in rodent malaria pathogenesis (Schofield *et al.* 2002). To prepare an immunogen, the synthetic GPI glycan was conjugated to the carrier protein keyhole limpet haemocyanin (KLH-glycan). Antibodies from mice immunized with KLH-glycan reacted only with malaria-infected rather than uninfected red blood cells, as demonstrated by immunofluorescence and Western blot analyses (Schofield *et al.* 2002). These results suggested that GPI of parasite origin was sufficiently different from host endogenous GPI molecules (which differ by virtue of amino-sugar or phosphoethanolamine modifications to the core glycan) to be recognized as foreign (non-self). These antibodies specifically neutralized the production of TNF-α from macrophages induced by *P. falciparum* extracts, confirming their anti-toxic properties *in vitro*. When challenged with *P. berghei* ANKA, naïve and sham-immunized mice all died within 5–8 days post infection (p.i.) with severe cerebral syndrome, whereas 75% of those mice immunized with the KLH-glycan were protected (Fig. 1a). There were no significant differences in parasite counts evident at this early stage of infection, but ultimately all mice succumbed at ~15 days p.i. to severe haemolytic anaemia (Fig. 1b) (Schofield *et al.* 2002) – indicating that parasite growth was unaffected by the antibodies. In studying the effectiveness of KLH-glycan vaccination, the immunized mice were found to be protected against occlusion of brain vasculature (Fig. 1c), pulmonary oedema (Fig. 1d) and blood acidosis (Fig. 1e) (Schofield *et al.* 2002). These findings not only demonstrated the efficacy of this prototype anti-toxic vaccine, but also showed that *Plasmodium* GPI is central to the pathogenesis of systemic disease endpoints and the development of a lethal cerebral syndrome in this model. Thus, GPI of *P. falciparum* satisfies the criteria for a predicted malaria toxin and may thereby contribute to life-threatening disease in humans, albeit a requirement in human malarial pathogenesis and fatality remains to be proven (Boutlis *et al.* 2002; Suguitan *et al.* 2004).

PLASMODIUM FALCIPARUM GPI DISTRIBUTION, METABOLISM AND STRUCTURE

GPI molecules represent a ubiquitous class of glycolipids that function as anchors for transmembrane proteins in eukaryotic cells. The parasitic protozoa are unusual in that they express very high levels of GPI-anchored proteins, GPI-anchored glycoconjugates and a range of unconjugated 'free' GPIs and GPI-like structures (GIPLs). Compared with mammalian cells, which typically express in the order of 10^5 GPI copies per cell, various parasitic protozoa express over 10^7 GPI copies per cell and most of these GPI-anchored molecules are displayed at the cell surface (Ferguson *et al.* 1994). In *P. falciparum*, which shows uniquely low levels of N- and O-linked glycosylation, GPI anchors constitute over 95% of the post-translational carbohydrate modification of parasite proteins (Davidson & Gowda, 2001; Gowda, Gupta & Davidson, 1997). Recent genome-wide sequence analyses predict that approximately 0·5% of the *P. falciparum* proteome may be post-translationally glycosylated with a GPI anchor, which is typical for a lower eukaryote (Eisenhaber, Bork & Eissnhaber, 2001). From microarray transcription data about 2/3 of these proteins are thought to be expressed in asexual blood stages (P. Gilson, personal communication). Among the GPI-anchored proteins in *Plasmodium*, the most important are the circumsporozoite protein (CS-2) which coats the insect stage sporozoites, and the merozoite surface proteins (e.g. MSP-1, MSP-2, MSP-4, MSP-5, MSP-10) expressed at the extracellular blood stage of the parasite. All of these proteins are of particular interest because of their central role in host/parasite interactions and in the response to host immunity, and several are under consideration as vaccine candidates. In addition to GPI-anchored proteins, *Plasmodium* species also contain an abundant pool of structurally identical GPI molecules which are not conjugated to proteins or other structures but exist free in the plasma membrane (free GPIs). In fact, quantification of the relative

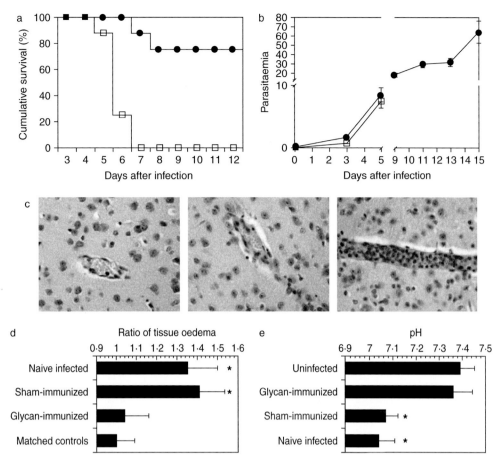

Fig. 1. Immunization against synthetic malaria GPI glycan substantially protects against murine cerebral malaria, pulmonary oedema and peripheral blood acidosis. (a), (b) Kaplan-Meier survival plots and parasitemia levels of KLH-glycan-immunized (filled circles) and sham-immunized (open squares) C57BL/6 mice challenged with *P. berghei* ANKA. (c) Haeomotoxylin and eosin-stained sections of brain tissue showing blood vessels from KLH-glycan immunized (left and centre panels) and sham-immunized (right panel) mice killed 6 days p.i. Pulmonary oedema and blood acidosis indices given as a proportion of the lung wet:dry weight ratio (d) or serum pH (e) form KLH-glycan-immunized and, sham-immunized and naïve infected mice at day 6 p.i. compared to age/sex-matched uninfected mice. Taken from Schofield *et al.* (1993). Permission pending.

amount of GPI molecules purified from late schizonts shows a 4-fold molar excess of free versus serine-linked GPIs (resulting from exhaustive pronase digestion of GPI-anchored proteins) by gas chromatography mass spectrometry (GC/MS), suggesting that free GPIs are the predominant GPI-anchored molecules present at the time of schizont rupture (K. Evans, personal communication). These free GPIs appear to be metabolic end-products and functionally important components in their own right.

Despite the diversity of GPI-anchored molecules found in parasitic protozoa, fungi and mammalian cells, all eukaryotic GPI membrane anchors have a common core structure, consisting of a conserved trimannosylglucosaminyl glycan (Man$_3$-GlN: Manα1,2-Manα1,6-Manα1,4-GlcN; Fig. 2, bold letters) linked to the inositol residue of a phosphatidylinositol (Man$_3$-GlN-PI). The glycan core of GPI anchors carries an ethanolamine phosphate (EtN-P) group on the third mannose (Etn-P-Man$_3$-GlN-PI), making it a substrate for GPI transamidase. This

enzyme cleaves the peptide bond at the GPI-anchor attachment site near the carboxy-terminal end and creates an amide linkage between the ethanolamine of GPI and the newly generated carboxyl group of the cleaved protein precursor. GPI molecules are assembled in the endoplasmic reticulum (ER) by the sequential addition of *N*-acetylglucosamine, core mannose and EtN-P residues to PI through the coordinated action of a series of glycsyl- and ethanolaminephosphate-transferases. Critical evaluation of sequence similarity data for malaria homologues of proteins essential for GPI synthesis indicates that the machinery for GPI assembly has been reduced to minimal requirements in *Plasmodium* (Delorenzi *et al.* 2002). Although the essential features of this pathway are conserved in all eukaryotes, considerable differences occur in the extent and nature of side chain additions to the glycan backbone as well as the nature of the lipid anchor (for excellent reviews on GPI structure and biosynthesis in parasitic protozoa see (McConville & Ferguson, 1993; McConville & Menon, 2000). As a result, the core glycan may be

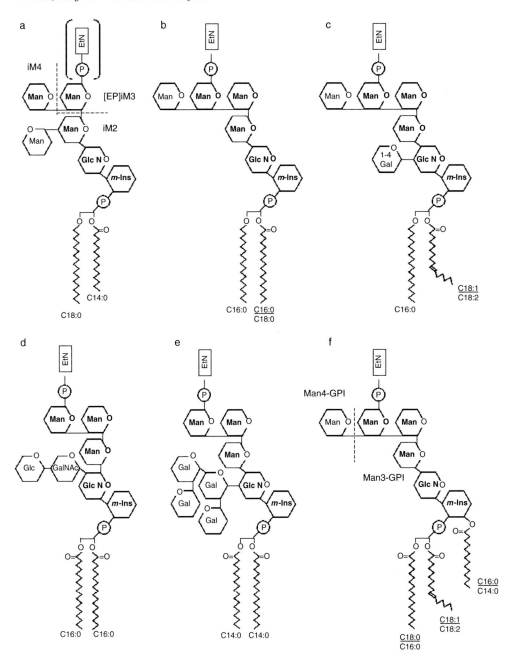

Fig. 2. Schematic structure of parasitic GPI-anchors: (a) *Leishmania mexicana* GIPLs (free iM2, iM3, [EP]iM3 and iM4 GPI forms); *Trypanosoma cruzi* epimastigote (b) and trypomastigote (c) mucin-derived GPI anchors; (d) *Toxoplasma gondii*-derived GPI anchor (e) *Trypanosoma brucei* VSG-derived GPI anchor (f) immature (Man$_3$-GPI) and mature MSP-1-derived (Man$_4$-GPI) GPI anchors of *Plasmodium falciparum*. Abbreviations: EtN, ethanolamine; P, phosphate; Man, mannose; GlcN, unacetylated glucosamine; Gal, Galactose; GalNAc, acetylated galactosamine, Glu, glucose; *m*-Ins, *myo*-Inositol; O, ether and ester bonded Oxygen molecules; Cn:n, number of saturated/ unsaturated carbon molecules in the lipid chain.

decorated with accessory carbohydrate (galactose, N-acetylglucosamine, sialic acid) side branches or EtN-P groups in a species- and tissue-specific manner (Fig. 2a-f). Other variations in the GPI structure are found in the lipid backbone (glycerol versus ceramide) and the number (mono-, di-), linkage (acyl versus alkyl), as well as the length and degree of saturation of the lipid chains (i. e. C14:0; C16:0, C18:0, C18:1, C18:2) in various GPIs (Fig. 2a–f).

The structure of *P. falciparum* GPIs was first characterized by Schwarz and colleagues (Gerold,

Diekmann-Schuppert & Schwartz, 1994), and is one of the simplest described to date (Fig. 2f). Analysis of radiolabeled GPIs suggested that the parasite synthesizes two mature GPI species, which were shown to correspond to a trimannosyl GPI (Fig 2f, Man$_3$-GPI) and a tetramannosyl GPI (Fig 2f, Man$_4$-GPI). The Man$_3$-GPI consists of the evolutionarily conserved core glycan linked to a diacylglycerol (EtN-P-Man$_3$-GlN-PI-DAG), whereas the Man$_4$-GPI carries a terminal fourth mannose residue in α1,2 linkage to the third mannose residue

(EtN-P-Man$_4$-GlN-PI-DAG). A report that the terminal fourth mannose residue carries an additional charged substituent (Gowda *et al.* 1997) remains unconfirmed. Detailed structural analysis by GC-MS has essentially confirmed the monosaccharide composition of the glycan moiety, but found considerable structural heterogeneity of native *P. falciparum* Man$_4$-GPI due to variation in fatty acid substituents of the lipid moiety (Naik *et al.* 2000). GC-MS analysis of HPLC-purified parasite GPIs treated with hydrofluoric acid or phospholipase A2 (for selective removal of fatty acids at the sn-1 and/or sn-2 position of the PI moiety, respectively) detected the presence of at least five distinct GPI diacylglycerol structures in the total free Man$_4$-GPI pool (Naik *et al.* 2000). These contained predominantly saturated stearic (C18:0) acid and unsaturated oleic (C18:1) acid esterified at the sn-1 and sn-2 positions, respectively, and had mainly palmitic (C16:0) acid on C2 of the inositol ring of the core glycan (Naik *et al.* 2000). The differences in fatty acid composition compared with earlier reports may reflect the preferential incorporation of various fatty acids present in different growth media into early GPI precursors (Mitamura *et al.* 2000; Krishnegowda & Gowda, 2003). Importantly, parasite proteins were found to be anchored almost exclusively by the Man$_4$-GPIs (Naik *et al.* 2000; Gerold *et al.* 1994), indicating a high degree of selectivity towards the presence of the fourth, terminal mannose residue in the GPI anchor moiety. By contrast, all five Man$_4$-GPI structures were found to be used for protein anchoring (Naik *et al.* 2000), irrespective of their lipid composition. Overall, *Plasmodium* GPIs therefore appear to represent an evolutionarily conserved glycolipid structure with a high level of variation in the lipid moiety (the precursors to GPI synthesis) and a structural requirement for the mature Man$_4$-GPI species (the substrate of GPI transamidase) with regard to protein anchoring.

STRUCTURE: ACTIVITY

The studies described in the previous sections demonstrated that *Plasmodium*-derived GPIs display potent pro-inflammatory activities that contribute to parasite-associated pathogenesis. Understanding the molecular details of the GPI structure-activity relationship and cell signalling mechanisms involved in these processes may therefore offer valuable insights that could lead to the development of novel therapeutics for malaria. Initial work on murine macrophages and human endothelial cells demonstrated that GPIs of *P. falciparum* and *T. brucei* origin can impart at least two discrete signals through structurally distinct elements. The core GPI glycan moiety was found to act as an agonist for the activation of protein tyrosine kinase (PTK), whereas the GPI diacylglycerol moiety was shown to function as

a second messenger substrate required for protein kinase C (PKC) activation (Tachado *et al.* 1997). Of the seven or so PKC members detectable in macrophages, activation by GPI diacylglycerols was restricted to the ε isoform. Activation was associated with a translocation from the cytoplasm to detergent-insoluble membranes. Transient tyrosine phosphorylation of *src*-family non-receptor tyrosine kinases (e.g. hck, lyn), which peaks within minutes after the addition of low concentrations (10 nM – 1 μM) of GPI to cells, is the earliest measurable event in GPI-mediated signal transduction (Tachado *et al.* 1997). Although *lyso*-GPI and deacylated inositolglycan generated by specific enzymatic degradations (i.e. PLA$_2$, PLD) were shown to be sufficient to activate PTKs, they were found not to promote the full range of signalling events required for the activation of macrophages. In addition, inhibition of signal transduction by specific PKC and NF-κB antagonists could block the production of cytokines in response to *P. falciparum*-derived GPIs (Tachado *et al.* 1997). Taken together, these data suggested that GPI-mediated PTK and PKC signals synergise to activate transcription factors from the NF-κB family for the induction of *de novo* expression of pro-inflammatory loci (e.g. TNF-α, IL-1, IL-6, iNOS, ICAM-1).

One of the important questions raised by these studies relates to the fine structure requirements with respect to the bioactivity of *P. falciparum* GPIs on host cells. Using structurally defined *Plasmodium* GPI precursors which differ in the degree of glycosylation together with specific enzymatic and chemical treatments for the selective removal of diacylglycerol or PI-diacylglycerol structures, it was possible to map the minimum structural requirement for PTK activation to the evolutionarily conserved Man$_3$-GlcN-PI moiety (Tachado *et al.* 1997). Consistent with this result, the biological activity of total parasite extract, including TNF-α output, can be inhibited by high concentrations of GPI partial structures such as α-methyl mannoside, glucosamine, and phosphatidylinositol. Vijaykumar *et al.* essentially confirmed and extended these findings by studying the inhibition profile of *P. falciparum* GPI bioactivity (Vijaykumar *et al.* 2001). In this study, authors used chemically defined GPI fragments generated by enzymatic treatment of HPLC-purified Man$_4$-GPI with α-mannosidase (for selective removal of the terminal fourth mannose residue on the core glycan) or chemical treatment with hydrofluoric acid or nitrous acid (for selective cleavage of phosphate or amino bonds of the core glycan, respectively). Consistent with results by Tachado *et al.* (1997), none of the GPI partial structures alone were able to induce TNF-α production by macrophages (Vijaykumar *et al.* 2001). Prior incubation of macrophages with inactive Man$_4$-containing carbohydrate fragments (but not the PI or DAG fragments) released by HF or nitrous acid was found to effectively inhibit the activity of intact Man$_4$-GPI (Vijaykumar *et al.* 2001). In contrast to intact Man$_4$-GPI, the authors reported that *P. falciparum* Man$_3$-GPI failed to activate macrophages, nor did Man$_3$-containing carbohydrate fragments inhibit the ability of Man$_4$-GPI to induce TNF-α production (Vijaykumar *et al.* 2001). Surprisingly, the same group subsequently published a paper reporting 80% of TNF-α output from macrophages in response to Man$_3$-GPI compared to Man$_4$-GPI (Krishnegowda *et al.* 2005). The data are therefore inconsistent and at odds with the different inhibition profile of soluble Man$_4$-containing and Man$_3$-containing carbohydrate fragments on *P. falciparum* GPI activity (Vijaykumar *et al.* 2001). Although discrepancies with regard to the bioactivity of intact Man$_3$-GPI remain to be resolved, current observations indicate that the fourth, terminal α-mannose residue linked to the core glycan moiety may be critical for the activity of *P. falciparum* GPIs.

The precise structural requirement of the lipid moiety for the bioactivity of *P. falciparum* GPIs still remains to be determined. Structure-activity data from different protozoan parasites indicates that ceramide or saturated lipid chains in the monoalkyl or 1-alkyl-2-acyl-glycerol GPIs moieties of *Leishmania* spp (Fig. 2a) or *T. cruzi* epimastigote forms (eGPI, Fig. 2b) induce relatively low or no activity in macrophages (Camargo *et al.* 1997*b*), compared to the bioactive *T. gondii* (Fig. 2d), *T. brucei* (Fig. 2e) and *P. falciparum* (Fig. 2f) diacylglycerol-containing GPI structures. Moreover, the bioactive GPIs of *T. cruzi* trypomastigote forms (tGPI, Fig. 2c) and *P. falciparum* GPI anchors (Fig. 2f) contain mainly unsaturated fatty acid chains at the sn-2 position in the 1-alkyl-2-acyl-glycerol and 1,2-diacylglycerol moieties, respectively (reviewed by (Almeida & Gazzinelli, 2001; Ropert *et al.* 2002). Whether the presence of the unsaturated fatty acid chain at the sn-2 position of *P. falciparum* GPI is actually required for its bioactivity is however unclear. Tachado *et al.* (1997) found that the potency of phospholipase A$_2$-generated sn-2 lyso GPIs of *P. falciparum* is drastically reduced, whereas Vijaykumar *et al.* (2001) and Krishnegowda *et al.* (2005) reported that these molecules retained nearly all activity. The latter findings are at odds with the observation that chemical removal of fatty acids at the sn-2 position is sufficient for a drastic ($\sim$100fold) reduction in the ability of highly purified *T. brucei* and *T. cruzi* trypomastigote GPIs to activate macrophages (Camargo *et al.* 1997*b*; Tachado *et al.* 1997, Almeida *et al.* 2000). The reason for this discrepancy is not clear, but is likely to be due to the difficulty in obtaining adequate amounts of pure, structurally-defined native GPIs. The preparation of native malarial GPI is a labour-intensive process that requires the large-scale growth of *P. falciparum* in

human red blood cell cultures. All growth media used must be endotoxin-free and cultures need to be rigorously tested for mycoplasma contamination before being processed, to eliminate the likelihood of potential contamination by bacterial lipopolysaccharides and mycoplasma-derived lipopeptides. The preparation of malarial GPI itself involves a series of optimized glycolipid extraction and fractionation steps using organic solvents and at least one round of RP-HPLC followed by analytical HP-TLC and GC-MS analyses. The yield of pure native GPI resulting from this procedure is typically low – in the range of 10 μg per 1 litre of culture – and this has proven to be probably one of the most limiting factors in the laboratory setting. The chemical synthesis of partial *Plasmodium* GPI structures (Schofield *et al.* 2002) as well as entire protozoan GPI-anchors (Debierre-Grockiego *et al.* 2003) may soon help to resolve some of the questions regarding the detailed structural requirements for malarial GPI bioactivity.

SIGNALLING PATHWAYS TRIGGERED BY *PLASMODIUM* GPI – THE TWO-SIGNAL MODEL

The molecular mechanisms of protozoan glycolipid-host cell interaction are poorly understood. In particular, how activation signals mediated by protozoan GPIs are transmitted across the plasma membrane to the interior of host immune cells remains unclear. Tachado *et al.* (1997) proposed a working hypothesis to explain how bioactive GPIs of *T. brucei* and *P. falciparum* can activate host macrophages. The initial interaction is predicted to involve the binding of parasite-derived glycolipids to a putative glycan-specific transmembrane receptor present in the surface of host macrophages. This agrees with the observation that stimulation of macrophages with parasite GPI glycan fragments alone is sufficient to cause rapid transient activation of *src*-family non-receptor PTKs in host macrophages. Although PTK activation does not provide the full range of signalling events leading to downstream inflammatory gene expression, capture by a glycan-specific receptor is predicted to catalyze the transfer of GPI from the extracellular space to the host membrane. Within the plasma membrane, GPI anchored molecules are known to diffuse laterally and partition into sphingolipid- and cholesterol-rich liquid-ordered microdomains which function both as signalling platforms and endosomal entry and sorting sites. Membrane-permeant diacylglycerol, formed via the hydrolysis of parasite-derived GPI molecules by PLD present in the serum or on the surface of mammalian macrophages (Krishnegowda *et al.* 2005), could cross cellular membranes to activate an intracellular PKC. Both PTK and PKC signalling pathways were shown to collaborate to activate transcription factors of the NF-κB/rel family and thereby promote expression of pro-inflammatory cytokines. Although certain aspects of this model remain unclear, it has provided a useful hypothetical framework that agrees with the observed structural requirement for both carbohydrate and fatty acid moieties of intact GPI-anchors of *P. falciparum*, *T. brucei*, *T. cruzi* and *T. gondii* origin for their full pro-inflammatory activity on macrophages (Almeida *et al.* 2000; Sileghem *et al.* 2001; Debierre-Grockiego *et al.* 2003). The major unresolved issues concerning this two-signal model for GPI-specific signal transduction are the identities of the glycan-specific transmembrane receptor and the molecular mechanism by which GPI-derived lipids contribute to the signalling pathways triggered by parasite-derived GPIs.

Toll-like receptor signalling pathways activated by parasite glycolipids

Recent efforts to identify the parasite glycolipid-specific receptor responsible for GPI-mediated transmembrane signalling have focused predominantly on certain members of the evolutionary conserved Toll-like receptor family. TLRs have emerged as key receptors responsible for recognizing conserved pathogen-associated molecular patterns (PAMPs) on microbes including modified lipids (e.g. bacterial lipopolysaccharides and lipoproteins), carbohydrates (e.g. yeast zymosan), proteins (e.g. flagellin), and nucleic acids (e.g. unmethylated CpG DNA and double-stranded RNA) (reviewed by Takeda, Kaisho & Akira, 2003; Akira & Takeda, 2004). The ten members of the human TLR family of receptors (TLRs 1-10) can form homo- or heterodimers with other TLR members or specific accessory proteins like CD14 to form a high-affinity binding site for multiple diverse ligands. This dimerisation triggers association with a family of adaptor molecules including MyD88, Mal/TIRAP and TRIF (O'Neill, 2003; McGettrick & O'Neill, 2004) that mediate downstream activation of IL-1R-associated kinases (IRAKs) and mitogen-activated protein kinases (MAPKs) that lead to the activation of NF-κB-mediated transcription of inflammatory cytokine genes.

The proposed involvement of different members of TLRs in cell signalling by parasite glycolipids was based on the observation that the pattern of macrophage activation by *T. cruzi* mucin-derived GPI-anchors is analogous to that of bacterial lipopolysacharide (LPS). Initial studies showed that, trypomastigote mucin GPIs are capable of triggering different MAP kinases (e.g. ERK1-2, MKK4 and p38) and NF-κB with kinetic and inhibition profiles similar to those triggered by LPS (Ropert *et al.* 2001). It was further demonstrated that an efficient response to bacterial lipopolysaccharide *in vivo* requires the formation of non-covalent CD14/LPS receptor interactions with the TLR-4/MD2 transmembrane receptor complex (Wright *et al.* 1990;

Poltorak *et al.* 1998; Nagai *et al.* 2002). However, macrophages from LPS-hypo-responsive C3H/HeJ mice with mutations in the TLR-4 receptor were still responsive to *T. cruzi*-derived GPI anchors or GPI-mucins (Ropert *et al.* 2001) – suggesting that these glycolipids trigger another member of the TLR family. Subsequent *in vitro* studies in mammalian cell lines transfected with different TLR expression plasmids and NF-κB-specific reporter constructs suggested that *T. cruzi* parasites GPI activate TLR-2 (Campos *et al.* 2001). Interestingly, GPI anchors derived from trypomastigote and epimastigote stages had variable activities in activating NF-κB-dependent gene expression in cell lines expressing CD14 and TLR-2 (Campos *et al.* 2001) – depending on the presence of extra galactose residues in the glycan core and unsaturated fatty acids in the sn-2 position of the alkylacylglycerolipid moiety for maximum activity. This agrees with the different potency of tGPIs and eGPIs in mouse macrophages and suggested that *T. cruzi* GPI-activation of TLR-2 is essential for the production of proinflammatory cytokines *in vitro*.

However, *in vivo* studies in TLR and MyD88 knockout mice in the susceptible C57BL/6 background have provided a rather more complex picture of the role of TLR-mediated signalling during acute *T. cruzi* infection. For example, when exposed to live *T. cruzi* trypomastigotes *in vitro*, macrophages from MyD88-deficient mice responded to neither purified trypomastitote GPIs nor live *T. cruzi* parasites (Campos *et al.* 2004). By contrast, macrophages from TLR-2 null mice, despite being less responsive to trypomastigote GPIs, were found to still produce pro-inflammatory cytokines when stimulated with live parasites (Campos *et al.* 2004). Moreover, whereas both MyD88 and IFN-γ knockout mice showed signs of enhanced susceptibility to *T. cruzi*–mediated lethality, TLR-2-deficient mice were almost as resistant to infection as wild-types (Campos *et al.* 2004). Taken together, the results indicate that the deficiency of a single TLR, such as TLR2, may not have a major impact during acute *T. cruzi* infection *in vivo*. This implies the existence of multiple TLR-dependent signalling pathways being activated during acute infection with live *T. cruzi* trypomastigotes, involving both mucin GPI anchors and probably other parasite-derived bioactive molecules. Consistent with similar findings in *Toxoplasma gondii* (Chen *et al.* 2002; Scanga *et al.* 2002) and *Leishmania major* (Hawn *et al.* 2002; De Veer *et al.* 2003; Muraille *et al.* 2003) infections, these data for the first time implied a major role for a TLR/MyD88-dependent signalling pathway in the early production of pro-inflammatory cytokines and host resistance to different parasites.

Current evidence for the involvement of TLR/MyD88-dependent mechanisms in the immunoregulation and immunopathogenesis of *Plasmodium* malaria is still sparse. A study in the BALB/c mouse model of *P. berghei* liver-stage infection demonstrated that cytokine-mediated liver injury depends on the TLR/MyD88 signalling pathway. This study found that both IL-12 and MyD88-deficient BALB/c mice infected with *P. berghei* parasites were unaffected by liver damage with reduced production of IL-12 but not IL-18 (Adachi *et al.* 2001). *P. berghei* infection therefore appears to activate a MyD88-dependent pathway to specifically induce IL-12, which in turn results in cytotoxic T cell lysis of hepatocytes and liver injury in experimental animals. However, none of the examined Toll-receptor-deficient strains (TLR-1, TLR-2, TLR-4, TLR-6 or TLR2/4 double-knockout mice) were found to affect IL-12 production or protect from liver damage (Adachi *et al.* 2001). The particular TLR involved in this process remains to be elucidated. It should be noted that cytokine-induced liver injury in this experimental system does not appear to model any known pathophysiological disease processes occurring in human malaria infections.

In contrast to malarial liver stages, which are clinically silent and short lived, severe malarial pathogenesis is largely mediated by the *Plasmodium* asexual blood stages. Therefore, it is of great interest to investigate the role of the TLR/MyD88 pathways during *P. berghei* ANKA-mediated malaria in the C57BL/6 strain susceptible to cerebral syndromes. A recent study reported that both human and murine plasmacytoid dendritic cells (pDCs) are activated by *Plasmodium* schizonts via a TLR-9/MyD88-dependent mechanism (Pichyangkul *et al.* 2004) to produce IFN-α. Coban *et al.* (2005) went on to show that haemozoin, a crystalline by-product derived from the digestion of erythrocytic haemoglobin is responsible for the activation of TLR-9 by malarial extracts. Members of the TLR-7 and TLR-9 subfamily are known to sense viral and bacterial CpG-DNA at the endosomal subcellular compartment via a strictly MyD88-dependent signalling pathway (Latz *et al.* 2004; Lund *et al.* 2004). Pro-inflammatory cytokine production, including TNF-α and IL-12p40 from Flt ligand-derived dendritic cells treated with malarial extracts is totally dependent on TLR-9 and MyD88 and independent of TRIF, another TLR adaptor involved in MyD88 independent TLR signalling (Coban *et al.* 2005). Production of IL-6 in sera of mice injected with purified haemozoin was dependent on MyD88 and TLR-9 (Coban *et al.* 2005) – suggesting the *in vivo* response to haemozoin is also MyD88 dependent (Coban *et al.* 2005). It should be noted however that haemozoin preparations vary in purity and may also be associated with bioactive contaminants.

While haemozoin is a TLR 9 ligand little else was known about other TLR activating components present in malarial extracts. A recent publication from Krishnegowda *et al.* (2005) has shown that purified *P. falciparum* GPI coupled to gold particles efficiently activates *in vitro*-derived bone marrow macrophages to produce pro-inflammatory cytokines in a MyD88 dependent manner. TNF-α production was shown to be 80% and 20% of that produced in macrophages from wild type mice when macrophages were derived from TLR4 and TLR2-null mice respectively. No significant activation of TNF-α was detected in macrophages isolated from either mice deficient in both TLR2/TLR4 or MyD88-null mice. This suggests that malarial GPI signalling in isolated bone marrow derived macrophages is primarily dependent on TLR2 and absolutely requires MyD88. Additional studies in HEK-293 cells transfected with different TLR expression plasmids and dual luciferase reporter constructs suggest that TLR2/TLR1 and TLR2/TLR5 heterodimers can differentially recognize GPIs containing three and two fatty acid substituents (Krishnegowda *et al.* 2005), in a manner similar to the discrimination of triacylated bacterial lipoproteins and diacylated mycoplasma lipoproteins (Takeda, Takeuchi & Akira, 2002). It would have been of interest to know the effects of other TLR-activating protozoal GPIs coupled to gold particles, as well as non TLR-activating glycolipids as a more relevant control than just gold particles alone. Importantly, whether TLR activation by *Plasmodium* GPI is a significant factor contributing to the pathogenesis of severe malaria syndromes *in vivo*, as implied by these studies, remains to be addressed.

It appears that GPI is important in the pathogenic cascade initiated by malarial infection as vaccination with the glycan moiety dramatically protects mice from cerebral symptoms (Schofield *et al.* 2002). Whether this pathogenic activity is mediated by TLR activated pathways or other mechanisms, is a crucial issue that remains to be established. To help address whether TLRs play a role in malarial pathogenesis we infected mice deficient in MyD88, which are unresponsive to all known TLR-2 and TLR-9 stimuli (McGettrick & O'Neill, 2004) including all the assays performed with malarial GPI (Krishnegowda *et al.* 2005) and haemozoin (Coban *et al.* 2005). Preliminary data indicate that while these mice fail to induce IL-12p40 in response to infection there is no significant difference in peak parasitaemia or death rates in MyD88$^{-/-}$ null mice compared to MyD88$^{+/-}$ heterozygous littermates (unpublished observation). This is in stark contrast to the protective effect observed in both IFN-γ and IFN-γ receptor knockout mice (Amani *et al.* 2000) and in mice immunized with the glycan portion of malarial GPI (Schofield *et al.* 2002). These data suggest that MyD88-independent pathways are the primary effectors involved in the initiation and amplification of the acute cerebral syndrome that is the pathogenic mode of action in this model of cerebral malaria.

A similar MyD88-independent mechanism appears to be responsible for the induction of IL-18 after *P. berghei*-infection of the resistant BALB/c strain (Adachi *et al.* 2001). Indeed, there are at least four additional MyD88-like adaptor molecules that may associate with TLRs to induced systemic inflammation (e.g. TIRAP/Mal, TRIF/TICAM-1 and TRAM/TICAM-2) (Horng *et al.* 2002; Yamamoto *et al.* 2002, 2003 *a, b*). Together, they are capable of providing specificity for MyD88-independent signalling by TLRs 1-4 and 6, but whether *Plasmodium* schizonts or purified glycolipids activate any of these TLRs is currently unknown. The activation of NKT-cells is important in the pathogenesis of rodent malaria (Hansen *et al.* 2003) and this offers another TLR-independent mechanism for the production of IFN-γ and initiation of the cascade that ultimately can lead to cerebral syndrome in the *P. berghei* model of malaria. Further studies will be required to dissect the potential role of *P. falciparum* GPI in the activation of multiple MyD88-dependent and -independent pathways in *P. berghei*-infected C57BL/6 mice and to elucidate the possible contribution of TLRs to severe malarial pathogenesis.

RECOGNITION OF PARASITE GPIS BY NON-TLR PATTERN RECOGNITION RECEPTORS

In addition to Toll-like receptors, cells of the immune system are equipped with many other glycan-specific surface proteins that function as pattern recognition and uptake receptors on macrophages and DCs. Many of these are members of the Ca^{2+}-dependent C-type lectin family and recognize their ligands through the structurally related Ca^{2+}-dependent carbohydrate-recognition domains (CRDs) (Cambi & Figdor, 2003). Pathogen recognition by lectin-like receptors (LLRs), such as the soluble serum mannose-binding protein (MBP) (Gadjeva, Takahashi & Tniel, 2004), the macrophage cell-surface mannose receptor (MMR) (Ezekowitz *et al.* 1990) or the dendritic cell-specific ICAM3-grabbing non-integrin (DC-SIGN) receptor (Cambi *et al.* 2003), is mediated by binding of terminal mannose residues characteristic of viral, bacterial, fungal and parasite cell surfaces. Ligand binding most likely depends on subtle differences in the arrangements of carbohydrate residues and their branching – creating unique sets of carbohydrate recognition profiles.

Current observations support the idea of carbohydrate-specific interactions in the recognition of protozoan glycolipids by macrophages. It has been suggested, for example, that the potent cytokine producing activity of *T. cruzi* mucin GPIs purified from trypomastigote forms may be due to the precise number and/or structure of carbohydrate residues, since mild treatment with periodate largely destroys this activity (Camargo *et al.* 1997 *a, b*). Compared to the low potency of *T. cruzi* mucin GPIs purified

from epimastigote forms the high potency of trypomastigote GPIs is probably due to up to four extra α-galactosyl residues linked to the first mannose residue of the glycan core (Almeida & Gazzinelli, 2001). In agreement with this hypothesis, the presence of the galactose side-chain additions to the core glycan of highly purified GPI anchors of *T. brucei* origin were also shown to be essential for optimal TNF-α output from macrophages (Magez *et al.* 1998). It is possible that this carbohydrate pattern is recognized by TLR2, as evidenced by the potent induction of NFκB-dependent gene expression in transfected cell lines by trypomastigote compared to epimastigote GPIs (Campos *et al.* 2001). Recent work by Vijaykumar *et al.* (2001) suggested that cell signalling by *P. falciparum* GPI molecules, which lack the galactosyl side chain modifications of *T. cruzi* and *T. brucei* GPIs, may represent a novel mechanism involving the initial recognition of the distal fourth α-mannosyl residue. This proposal is based on the carbohydrate inhibition profile of *P. falciparum* Man_4-GPI activity on macrophages *in vitro*, as discussed above.

A good example of a recognition molecule able to differentiate between defined mannose-terminating patterns is the hepatocyte-derived soluble mannan-binding protein. MBP is an abundant C-type lectin found in various mammalian sera including human sera, and plays an important role in innate immunity. It is able to specifically recognize terminal mannose-, *N*-acetyoglucosamine and fucose-rich carbohydrate patterns on the surfaces of pathogenic microorganisms and initiates complement deposition on relevant (non-self) surfaces (reviewed by Gadjeva *et al.* 2004). In addition to carbohydrate patterns present on bacteria, viruses and fungi, it has been observed that MBP is also able to recognize major surface glycoconjugates and glycoproteins of *Leishmania* and *P. falciparum* parasites. For example, using thin layer chromatogram-overlay assays it was shown that MBP binds to the mannose-terminating di-, tri- and tetrasaccharide fragments derived from purified LPG and GIPLs that are highly expressed on promastigote and amastigote stages of most *Leishmania* species (Green *et al.* 1994). As demonstrated by immunofluorescence micrococy of live *L. major* and *L. mexicana* promastigotes (Green *et al.* 1994), FACS analyses show that human MBL may also function as an opsonin for schizont-infected erythrocytes (Klabunde *et al.* 2002). Attempts to identify specific parasite-derived surface ligands found that MBP binds to at least two glycoproteins that incorporated ^{3}H-glucosamine and were immunogenic in humans (Klabunde *et al.* 2002; Garred *et al.* 2003). In patients with homozygous mutations in the MBL gene complicated malaria was associated with significantly higher parasite burdens compared to MBL-competent counterparts (Garred *et al.* 2003), but whether MBP deficiency is a significant risk factor for severe malaria is still a matter of debate (Bellamy *et al.* 1998).

Whether tissue macrophages or dendritic cells express a specific lectin-like receptor, such as cell-surface mannose receptor or DC-SIGN capable of recognizing the terminal mannose residue on the core glycan of *Plasmodium* GPIs remains to be determined. Evidence is emerging that LLRs not only play a role as phagocytic antigen uptake receptors, but may allow signalling upon fusion of late endosomes/lysosomes via TLRs expressed on endosomal membranes (i.e. TLR-3, TLR-7, TLR-9). Whereas TLR signalling through MyD88, IRAK and MAPK resulting in the activation of NF-κB is quite well understood, signalling through lectin-like receptors via alternative pathways is still relatively obscure. Nevertheless, in view of the different dose-response and TLR/MyD88 signalling profiles of *T. cruzi* trypomastigote and *P. falciparum* GPIs, it appears that cellular immune responses to GPIs of various parasitic protozoa may involve multiple lipid- and carbohydrate-specific pattern recognition and signal transduction mechanisms.

SIGNALLING BY GPI-ANCHORED PROTEINS AND GLYOLIPIDS THROUGH LIPID-DEPENDENT MICRODOMAINS

A striking property of mammalian GPI-anchored proteins as well as glycolipids is that their ligation on the cell surface by natural ligands or suitable antibodies regulates numerous signal transduction and activation responses (reviewed by (Horejsi *et al.* 1998; Kasahara & Sanai, 2000). Such signalling capacity is surprising, considering that these molecules have no transmembrane or intracellular moieties and thus no direct contact with the cell interior. The signalling capacity of both endogenously expressed and exogenously added GPI-anchored proteins appears in many cases to involve their association with cytoplasmic signalling molecules in detergent-resistant membrane domains (DRMs). DRMs are stabilized against disruption by cold detergents through liquid-ordered packing of component cholesterol, (glycol)sphingolipids and glycerolipids. An important factor in this lipid-dependent organisation appears to be the length and fluidity (saturation) of the fatty acid residues attached with the GPI-anchored molecules (Schroeder *et al.* 1998; Benting *et al.* 1999). Low buoyant density DRMs, often called lipid rafts, can be isolated by density gradient ultracentrifugation. Such membrane fractions are characteristically enriched in GPI-anchored proteins, dually acylated cytoplasmic signalling or adaptor proteins (e.g. Src-family kinases, trimeric G proteins, Ras, LAT) and different raft-organizing proteins (e.g. caveolins, stomatins, flotillins) (Brown & London, 2000). *In vivo*, liquid-ordered membrane domains are thought to be quite small and/or

extremely dynamic and consequently below the level of resolution of the light microscope (Jacobson & Dietrich, 1999). Thus, they are optimally investigated with biophysical techniques that can detect short-range interactions. Recent studies employing chemical crosslinking and fluorescence resonance energy transfer (FRET) microscopy techniques estimated that a fraction (20–40%) of GPI-anchored proteins form extremely tight clusters of nanometer size ($\sim$5–50 nm), each containing a few (4–15) different molecules on the surface of living cells (Friedrichson & Kurzchalia, 1998; Varma & Mayor, 1998; Sharma *et al.* 2004). Following cross-linking with antibodies or natural ligands these rafts can coalesce into visible and stable functional domains. Lipid-dependent membrane organization has been proposed to be responsible for the endocytosis and sorting of raft-associated proteins, and for providing platforms for the dynamic assembly of GPI-anchored proteins, transmembrane proteins and cytoplasmic signalling molecules in a variety of biological contexts (e.g. T cell receptor, B cell receptor and FcεRI signalling) (Simons & Toomre, 2000; Anderson & Jacobson, 2002; Mayor & Riezman, 2004).

Of particular interest in the present context of *Plasmodium* GPI signalling is accumulating evidence indicating that the innate recognition of different pathogens may involve the interaction of CD14 with various transmembrane receptors within supramolecular activation clusters (i.e. TLR-2, TLR-4, TLR-6, Mac-1, CXCR4, GDF5 and others) (Henneke *et al.* 2001; Pfeiffer *et al.* 2001; Triantafilou *et al.* 2001 *a, b*). By analogy to the immunological synapse, it has been proposed that the entire bacterial recognition system may be based around the formation of a signalling complex of receptors at the site of CD14-LPS ligation within membrane microdomains (Triantafilou & Triantafilou, 2002). Consistent with this theory, FRET analyses have shown that ligation of CD14 by LPS and ceramide provokes ligand-specific clustering and translocation of TLR4 into lipid rafts (Pfeiffer *et al.* 2001; Triantafilou *et al.* 2002). A recent study demonstrates that the atypical PKC isoform, PKC-ζ, is critical in the regulation of LPS-induced TLR4 lipid raft mobilization within macrophages (Cuschieri, Umanskiy & Solomkin, 2004). The cellular mechanisms involved in this process remain incompletely understood, but recruitment of PKCζ to plasma membrane lipid raft microdomains during insulin stimulation of adipocytes was shown to be mediated by TC10, a member of the Rho family of small GTP-binding proteins that is constitutively localized to plasma membrane lipid raft microdomains (Kanzaki *et al.* 2004).

It is possible that similar processes are required for signalling by biologically active parasite GPIs. We recently generated fluorescently-labelled semi-synthetic *P. falciparum* GPIs to study the specio-temporal distribution of exogenously added parasite glycolipids in the host cell membrane by confocal microscopy. When added to live mammalian tissue culture cells *in vitro*, these molecules were initially distributed homogeneously, but after several hours concentrated at cell-to-cell contacts and acquired a punctuate distribution on the cell surface (unpublished observation). These preliminary data demonstrate that a proportion of *Plasmodium* glycolipids incorporate into host plasma membrane, but whether the observed cell surface puncta represent sites involved in uptake or signalling by *Plasmodium* glycolipids is currently unknown. FRET analyses using fluorescently-tagged semi- or fully synthetic *Plasmodium* GPI molecules may allow us to further investigate the possible association of *P. falciparum* GPI with CD14, TLRs and other signalling molecules within plasma membrane microdomains in live cells.

CONCLUSIONS AND PERSPECTIVES

While our understanding of malarial GPI bioactivity has improved significantly over the past decade, several key issues remain to be resolved. For example, while the concept that the GPI anchors of parasitic protozoa constitute immunostimulatory and regulatory agents in the context of protozoan infections has been widely recognized, it is still difficult to pinpoint the precise molecular features of parasite GPIs that underlie the differences in their bioactivities. This is especially true in the case of malarial GPI toxin, with independent studies coming to different conclusions with regard to the structural requirement for the presence of the terminal fourth mannose residue on the core glycan and/or the unsaturated fatty acid chain at the sn-2 position of *P. falciparum* GPI for full activation of pro-inflammatory responses of macrophages. One of the main factors contributing to inconsistencies in the present data, and at the same time limiting the pace of our research, is the difficulty in obtaining adequate amounts of pure, structurally defined *P. falciparum* GPIs. The chemical synthesis of milligram quantities of partial GPI structures as well as the entire *Plasmodium* GPI-anchor will make this key reagent more accessible and facilitate the creation of novel probes that will help to resolve some of the questions regarding fine structural requirements for malarial GPI bioactivity, in the near future.

A significant development regarding the potential mechanism of GPI bioactivity has been the demonstration that GPI anchors of various parasitic protozoa can mediate cellular immune responses via the Toll-like receptor family. The original observation that *T. cruzi* trypomastigote GPIs can specifically activate TLR2/MyD88-dependent signalling pathways to produce inflammatory responses in host macrophages has recently been confirmed and extended using *P. falciparum* GPI. The use of

macrophages from mice deficient in individual or multiple TLRs (e.g. TLR-2, TLR-4, TLR-2 & 4, TLR-9 etc.), essential adaptor molecules (i.e. MyD88), and downstream signalling molecules (e.g. ERK1/2, JNK-1 or JNK-2) has allowed investigators to dissect the relative contributions of specific TLR(s) and downstream signalling pathways activated by *P. falciparum* GPI and other malarial ligands. Recent studies indicate that malarial GPIs and haemozoins are preferentially recognized by TLR2/TLR1 and TLR9, respectively. Both rely on the MyD88-dependent activation of specific signalling pathways that mediate the production of proinflammatory cytokines and nitric oxide. However, the contribution of bioactive malaria products to severe malarial pathology and the role of TLRs or other pattern recognition receptors in innate immunity is only just beginning to be characterized. Importantly, the role of TLR/MyD88 pathways in mediating pro-inflammatory cytokine production and clearance of blood-stage parasites during the acute phase of malaria infections in mice infected with various *Plasmodium* species, as well as the role of GPI as a critical toxin in the context of human malaria infections remains to be firmly established. Given the complexity of the host/pathogen interface, signalling of malarial GPIs via TLR/MyD88-dependent pathways most likely represents only one example of how multiple pattern recognition receptors may function in regulating innate and adaptive immune responses in the context of protozoal infections. Further analysis of the interaction of *P. falciparum* parasites and the host immune system may shed new insights into the molecular mechanisms underlying severe malarial pathogenesis with potentially important implications for the development of novel therapeutics for malaria treatment.

ACKNOWLEDGMENTS

Supported by the NH & MRC, NIH, HFSP and the UNDP/World Bank/WHO Program (TDR). L.S. is an International Research Scholar of the Howard Hughes Medical Institute.

REFERENCES

ADACHI, K., TSUTSUI, H., KASHIWAMURA, S., SEKI, E., NAKANO, H., TAKEUCHI, O., TAKEDA, K., OKUMURA, K., VAN KAER, L., OKAMURA, H., AKIRA, S. & NAKANISHI, K. (2001). *Plasmodium berghei* infection in mice induces liver injury by an IL-12- and toll-like receptor/myeloid differentiation factor 88-dependent mechanism. *Journal of Immunlogy* **167**, 5928–5934.

AKIRA, S. & TAKEDA, K. (2004). Toll-like receptor signalling. *Nature Reviews Immunology* **4**, 499–511.

AL YAMAN, F. M., MOKELA, D., GENTON, B., ROCKETT, K. A., ALPERS, M. P. & CLARK, I. A. (1996). Association between serum levels of reactive nitrogen intermediates and coma in children with cerebral malaria in Papua New Guinea.

Transaction of the Royal Society of Tropical Medicine and Hygiene **90**, 270–273.

ALMEIDA, I. C., CAMARGO, M. M., PROCOPIO, D. O., SILVA, L. S., MEHLERT, A., TRAVASSOS, L. R., GAZZINELLI, R. T. & FERGUSON, M. A. (2000). Highly purified glycosylphosphatidylinositols from *Trypanosoma cruzi* are potent proinflammatory agents. *EMBO Journal* **19**, 1476–1485.

ALMEIDA, I. C. & GAZZINELLI, R. T. (2001). Proinflammatory activity of glycosylphosphatidylinositol anchors derived from *Trypanosoma cruzi*: structural and functional analyses. *Journal of Leukocyte Biology* **70**, 467–477.

AMANI, V., VIGARIO, A. M., BELNOUE, E., MARUSSIG, M., FONSECA, L., MAZIER, D. & RENIA, L. (2000). Involvement of IFN-gamma receptor-medicated signaling in pathology and anti-malarial immunity induced by *Plasmodium berghei* infection. *European Journal of Immunology* **30**, 1646–1655.

ANDERSON, R. G. & JACOBSON, K. (2002). A role for lipid shells in targeting proteins to caveolae, rafts, and other lipid domains. *Science* **296**, 1821–1825.

BELLAMY, R., RUWENDE, C., MCADAM, K. P., THURSZ, M., SUMIYA, M., SUMMERFIELD, J., GILBERT, S. C., CORRAH, T., KWIATKOWSKI, D., WHITTLE, H. C. & HILL, A. V. (1998). Mannose binding protein deficiency is not associated with malaria, hepatitis B carriage nor tuberculosis in Africans. *Quarterly Journal of Medicine* **91**, 13–18.

BENTING, J., RIETVELD, A., ANSORGE, I. & SIMONS, K. (1999). Acyl and alkyl chain length of GPI-anchors is critical for raft association *in vitro*. *FEBS Letters* **462**, 47–50.

BERENDT, A. R., SIMMONS, D. L., TANSEY, J., NEWBOLD, C. I. & MARSH, K. (1989). Intercellular adhesion molecule-1 is an endothelial cell adhesion molecule for *Plasmodium falciparum*. *Nature* **341**, 57–59.

BOUTLIS, C. S., GOWDA, D. C., NAIK, R. S., MAGUIRE, G. P., MGONE, C. S., BOCKARIE, M. J., LAGOG, M., IBAM, E., LORRY, K. & ANSTEY, N. M. (2002). Antibodies to *Plasmodium falciparum* glycosylphosphatidylinositols: inverse association with tolerance of parasitemia in Papua New Guinean children and adults. *Infection and Immunity* **70**, 5052–5057.

BROWN, D. A. & LONDON, E. (2000). Structure and function of sphingolipid- and cholesterol-rich membrane rafts. *Journal of Biological Chemistry* **275**, 17221–17224.

CAMARGO, M. M., ALMEIDA, I. C., PEREIRA, M. E., FERGUSON, M. A., TRAVASSOS, L. R. & GAZZINELLI, R. T. (1997*a*). Glycosylphosphatidylinositol-anchored mucin-like glycoproteins isolated from *Trypanosoma cruzi* trypomastigotes initiate the synthesis of proinflammatory cytokines by macrophages. *Journal of Immunology* **158**, 5890–5901.

CAMARGO, M. M., ANDRADE, A. C., ALMEIDA, I. C., TRAVASSOS, L. R. & GAZZINELLI, R. T. (1997*b*). Glycoconjugates isolated from *Trypanosoma cruzi* but not from *Leishmania* species membranes trigger nitric oxide synthesis as well as microbicidal activity in IFN-gamma-primed macrophages. *Journal of Immunology* **159**, 6131–6139.

CAMBI, A. & FIGDOR, C. G. (2003). Dual function of C-type lectin-like receptors in the immune system. *Current Opinion in Cell Biology* **15**, 539–546.

CAMBI, A., GIJZEN, K., DE VRIES, J. M., TORENSMA, R., JOOSTEN, B., ADEMA, G. J., NETEA, M. G., KULLBERG, B. J., ROMANI, L. & FIGDOR, C. G. (2003). The C-type lectin

DC-SIGN (CD209) is an antigen-uptake receptor for *Candida albicans* on dendritic cells. *European Journal of Immunology* **33**, 532–538.

CAMPOS, M. A., ALMEIDA, I. C., TAKEUCHI, O., AKIRA, S., VALENTE, E. P., PROCOPIO, D. O., TRAVASSOS, L. R., SMITH, J. A., GOLENBOCK, D. T. & GAZZINELLI, R. T. (2001). Activation of Toll-like receptor-2 by glycosylphosphatidylinositol anchors from a protozoan parasite. *Journal of Immunology* **167**, 416–423.

CAMPOS, M. A., CLOSEL, M., VALENTE, E. P., CARDOSO, J. E., AKIRA, S., ALVAREZ-LEITE, J. I., ROPERT, C. & GAZZINELLI, R. T. (2004). Impaired production of proinflammatory cytokines and host resistance to acute infection with *Trypanosoma cruzi* in mice lacking functional myeloid differentiation factor 88. *Journal of Immunology* **172**, 1711–1718.

CHANG, W. L., JONES, S. P., LEFER, D. J., WELBOURNE, T., SUN, G., YIN, L., SUZUKI, H., HUANG, J., GRANGER, D. N. & VAN DER HEYDE, H. C. (2001). CD8(+)-T-cell depletion ameliorates circulatory shock in *Plasmodium berghei*-infected mice. *Infection and Immunity* **69**, 7341–7348.

CHEN, M., AOSAI, F., NOROSE, K., MUN, H. S., TAKEUCHI, O., AKIRA, S. & YANO, A. (2002). Involvement of MyD88 in host defense and the down-regulation of anti-heat shock protein 70 autoantibody formation by MyD88 in *Toxoplasma gondii*-infected mice. *Journal of Parasitology* **88**, 1017–1019.

CHEN, Q., SCHLICHTHERLE, M. & WAHLGREN, M. (2000). Molecular aspects of severe malaria. *Clinical Microbiology Reviews* **13**, 439–450.

COBAN, C., ISHII, K. J., KAWAI, T., HEMMI, H., SATO, S., UEMATSU, S., YAMAMOTO, M., TAKEUCHI, O., ITAGAKI, S., KUMAR, N., HORII, T. & AKIRA, S. (2005). Toll-like receptor 9 mediates innate immune activation by the malaria pigment hemozoin. *Journal of Experimental Medicine* **201**, 19–25.

CUSCHIERI, J., UMANSKIY, K. & SOLOMKIN, J. (2004). PKC-zeta is essential for endotoxin-induced macrophage activation. *Journal of Surgical Research* **121**, 76–83.

DAVIDSON, E. A. & GOWDA, D. C. (2001). Glycobiology of *Plasmodium falciparum*. *Biochimie* **83**, 601–604.

DE SOUZA, J. B. & RILEY, E. M. (2002). Cerebral malaria: the contribution of studies in animal models to our understanding of immunopathogenesis. *Microbes and Infection* **4**, 291–300.

DE VEER, M. J., CURTIS, J. M., BALDWIN, T. M., DIDONATO, J. A., SEXTON, A., McCONVILLE, M. J., HANDMAN, E. & SCHOFIELD, L. (2003). MyD88 is essential for clearance of *Leishmania major*: possible role for lipophosphoglycan and Toll-like receptor 2 signalling. *European Journal of Immunology* **33**, 2822–2831.

DEBIERRE-GROCKIEGO, F., AZZOUZ, N., SCHMIDT, J., DUBREMETZ, J. F., GEYER, H., GEYER, R., WEINGART, R., SCHMIDT, R. R. & SCHWARZ, R. T. (2003). Roles of glycosylphosphatidylinositols of *Toxoplasma gondii*. Induction of tumor necrosis factor-alpha production in macrophages. *Journal of Biological Chemistry* **278**, 32987–32993.

DELORENZI, M., SEXTON, A., SHAMS-ELDIN, H., SCHWARZ, R. T., SPEED, T. & SCHOFIELD, L. (2002). Genes for glyco-sylphosphatidylinositol toxin biosynthesis in *Plasmodium falciparum*. *Infection and Immunity* **70**, 4510–4522.

EISENHABER, B., BORK, P. & EISENHABER, F. (2001). Post-translational GPI lipid anchor modification of proteins in kingdoms of life: analysis of protein sequence data from complete genomes. *Protein Engineering* **14**, 17–25.

EZEKOWITZ, R. A., SASTRY, K., BAILLY, P. & WARNER, A. (1990). Molecular characterization of the human macrophage mannose receptor: demonstration of multiple carbohydrate recognition-like domains and phagocytosis of yeasts in Cos-1 cells. *Journal of Experimental Medicine* **172**, 1785–1794.

FERGUSON, M. A., BRIMACOMBE, J. S., COTTAZ, S., FIELD, R. A., GUTHER, L. S., HOMANS, S. W., McCONVILLE, M. J., MEHLERT, A., MILNE, K. G., RALTON, J. E. & *et al.* (1994). Glycosyl-phosphatidylinositol molecules of the parasite and the host. *Parasitology* **108** (Suppl.) S45–S54.

FRIEDRICHSON, T. & KURZCHALIA, T. V. (1998). Microdomains of GPI-anchored proteins in living cells revealed by crosslinking. *Nature* **394**, 802–805.

GADJEVA, M., TAKAHASHI, K. & THIEL, S. (2004). Mannan-binding lectin–a soluble pattern recognition molecule. *Molecular Immunology* **41**, 113–121.

GARRED, P., NIELSEN, M. A., KURTZHALS, J. A., MALHOTRA, R., MADSEN, H. O., GOKA, B. Q., AKANMORI, B. D., SIM, R. B. & HVIID, L. (2003). Mannose-binding lectin is a disease modifier in clinical malaria and may function as opsonin for *Plasmodium falciparum*-infected erythrocytes. *Infection and Immunity* **71**, 5245–5253.

GEROLD, P., DIECKMANN-SCHUPPERT, A. & SCHWARZ, R. T. (1994). Glycosylphosphatidylinositols synthesized by asexual erythrocytic stages of the malarial parasite, *Plasmodium falciparum*. Candidates for plasmodial glycosylphosphatidylinositol membrane anchor precursors and pathogenicity factors. *Journal of Biological Chemistry* **269**, 2597–2606.

GOWDA, D. C., GUPTA, P. & DAVIDSON, E. A. (1997). Glycosylphosphatidylinositol anchors represent the major carbohydrate modification in proteins of intraerythrocytic stage *Plasmodium falciparum*. *Journal of Biological Chemistry* **272**, 6428–6439.

GRAU, G. E., FAJARDO, L. F., PIGUET, P. F., ALLET, B., LAMBERT, .H. & VASSALLI, P. (1987). Tumor necrosis factor (cachectin) as an essential mediator in murine cerebral malaria. *Science* **237**, 1210–1212.

GRAU, G. E., HEREMANS, H., PIGUET, P. F., POINTAIRE, P., LAMBERT, P. H., BILLIAU, A. & VASSALLI, P. (1989*a*). Monoclonal antibody against interferon gamma can prevent experimental cerebral malaria and its associated overproduction of tumor necrosis factor. *Proceedings of the National Academy of Sciences, USA* **86**, 5572–5574.

GRAU, G. E., POINTAIRE, P., PIGUET, P. F., VESIN, C., ROSEN, H., STAMENKOVIC, I., TAKEI, F. & VASSALLI, P. (1991). Late administration of monoclonal antibody to leukocyte function-antigen 1 abrogates incipient murine cerebral malaria. *European Journal of Immunology* **21**, 2265–2267.

GRAU, G. E., TAYLOR, T. E., MOLYNEUX, M. E., WIRIMA, J. J., VASSALLI, P., HOMMEL, M. & LAMBERT, P. H. (1989*b*). Tumor necrosis factor and disease severity in children with falciparum malaria. *New England Journal of Medicine* **320**, 1586–1591.

GREEN, P. J., FEIZI, T., STOLL, M. S., THIEL, S., PRESCOTT, A. & McCONVILLE, M. J. (1994). Recognition of the major cell surface glycoconjugates of *Leishmania* parasites by the human serum mannan-binding protein. *Molecular and Biochemical Parasitology* **66**, 319–328.

HANSEN, D. S., SIOMOS, M. A., BUCKINGHAM, L., SCALZO, A. A. & SCHOFIELD, L. (2003). Regulation of murine cerebral malaria pathogenesis by CD1d-restricted NKT cells and the natural killer complex. *Immunity* **18**, 391–402.

HAWN, T. R., OZINSKY, A., UNDERHILL, D. M., BUCKNER, F. S., AKIRA, S. & ADEREM, A. (2002). *Leishmania major* activates IL-1 alpha expression in macrophages through a MyD88-dependent pathway. *Microbes and Infection* **4**, 763–771.

HENNEKE, P., TAKEUCHI, O., VAN STRIJP, J. A., GUTTORMSEN, H. K., SMITH, J. A., SCHROMM, A. B., ESPEVIK, T. A., AKIRA, S., NIZET, V., KASPER, D. L. & GOLENBOCK, D. T. (2001). Novel engagement of CD14 and multiple toll-like receptors by group B streptococci. *Journal of Immunology* **167**, 7069–7076.

HOREJSI, V., CEBECAUER, M., CERNY, J., BRDICKA, T., ANGELISOVA, P. & DRBAL, K. (1998). Signal transduction in leucocytes via GPI-anchored proteins: an experimental artefact or an aspect of immunoreceptor function? *Immunology Letters* **63**, 63–73.

HORNG, T., BARTON, G. M., FLAVELL, R. A. & MEDZHITOV, R. (2002). The adaptor molecule TIRAP provides signalling specificity for Toll-like receptors. *Nature* **420**, 329–333.

JACOBSON, K. & DIETRICH, C. (1999). Looking at lipid rafts? *Trends in Cell Biology* **9**, 87–91.

JENNINGS, V. M., ACTOR, J. K., LAL, A. A. & HUNTER, R. L. (1997). Cytokine profile suggesting that murine cerebral malaria is an encephalitis. *Infection and Immunity* **65**, 4883–4887.

KANZAKI, M., MORA, S., HWANG, J. B., SALTIEL, A. R. & PESSIN, J. E. (2004). Atypical protein kinase C (PKCzeta/lambda) is a convergent downstream target of the insulin-stimulated phosphatidylinositol 3-kinase and TC10 signaling pathways. *Journal of Cell Biology* **164**, 279–290.

KASAHARA, K. & SANAI, Y. (2000). Functional roles of glycosphingolipids in signal transduction via lipid rafts. *Glycoconjugates Journal* **17**, 153–162.

KLABUNDE, J., UHLEMANN, A. C., TEBO, A. E., KIMMEL, J., SCHWARZ, R. T., KREMSNER, P. G. & KUN, J. F. (2002). Recognition of *Plasmodium falciparum* proteins by mannan-binding lectin, a component of the human innate immune system. *Parasitology Research* **88**, 113–117.

KOCH, O., AWOMOYI, A., USEN, S., JALLOW, M., RICHARDSON, A., HULL, J., PINDER, M., NEWPORT, M. & KWIATKOWSKI, D. (2002). IFNGR1 gene promoter polymorphisms and susceptibility to cerebral malaria. *Journal of Infectious Diseases* **185**, 1684–1687.

KRISHNEGOWDA, G. & GOWDA, D. C. (2003). Intraerythrocytic *Plasmodium falciparum* incorporates extraneous fatty acids into its lipids without any structural modification. *Molecular and Biochemical Parasitology* **132**, 55–58.

KRISHNEGOWDA, G., HAJJAR, A. M., ZHU, J., DOUGLASS, E. J., UEMATSU, S., AKIRA, S., WOODS, A. S. & GOWDA, D. C. (2005). Induction of proinflammatory responses in macrophages by the glycosylphosphatidylinositols of *Plasmodium falciparum*: cell signalling receptors, glycosylphosphatidylinositol (GPI) structural requirement, and regulation of GPI activity. *Journal of Biological Chemistry* **280**, 8606–8616.

LATZ, E., SCHOENEMEYER, A., VISINTIN, A., FITZGERALD, K. A., MONKS, B. G., KNETTER, C. F., LIEN, E., NILSEN, N. J., ESPEVIK, T. & GOLENBOCK, D. T. (2004). TLR9 signals after translocating from the ER to CpG DNA in the lysosome. *Nature Immunology* **5**, 190–198.

LUCAS, R., JUILLARD, P., DECOSTER, E., REDARD, M., BURGER, D., DONATI, Y., GIROUD, C., MONSO-HINARD, C., DE KESEL, T., BUURMAN, W. A., MOORE, M. W., DAYER, J. M., FIERS, W., BLUETHMANN, H. & GRAU, G. E. (1997). Crucial role of tumor necrosis factor (TNF) receptor 2 and membrane-bound TNF in experimental cerebral malaria. *European Journal of Immunology* **27**, 1719–1725.

LUND, J. M., ALEXOPOULOU, L., SATO, A., KAROW, M., ADAMS, N. C., GALE, N. W., IWASAKI, A. & FLAVELL, R. A. (2004). Recognition of single-stranded RNA viruses by Toll-like receptor 7. *Proceedings of the National Academy Sciences, USA*, **101**, 5598–5603.

LYKE, K. E., BURGES, R., CISSOKO, Y., SANGARE, L., DAO, M., DIARRA, I., KONE, A., HARLEY, R., PLOWE, C. V., DOUMBO, O. K. & SZTEIN, M. B. (2004). Serum levels of the proinflammatory cytokines interleukin-1 beta (IL-1beta), IL-6, IL-8, IL-10, tumor necrosis factor alpha, and IL-12(p70) in Malian children with severe *Plasmodium falciparum* malaria and matched uncomplicated malaria or healthy controls. *Infection and Immunity* **72**, 5630–5637.

MAGEZ, S., STIJLEMANS, B., RADWANSKA, M., PAYS, E., FERGUSON, M. A. & DE BAETSELIER, P. (1998). The glycosyl-inositol-phosphate and dimyristoylglycerol moieties of the glycosylphosphatidylinositol anchor of the trypanosome variant-specific surface glycoprotein are distinct macrophage-activating factors. *Journal of Immunology* **160**, 1949–1956.

MAYOR, S. & RIEZMAN, H. (2004). Sorting GPI-anchored proteins. *Nature Reviews: Molecular Cell Biology* **5**, 110–120.

McCONVILLE, M. J. & FERGUSON, M. A. (1993). The structure, biosynthesis and function of glycosylated phosphatidylinositols in the parasitic protozoa and higher eukaryotes. *Biochemical Journal* **294**, 305–324.

McCONVILLE, M. J. & MENON, A. K. (2000). Recent developments in the cell biology and biochemistry of glycosylphosphatidylinositol lipids (review). *Molecular Membrane Biology* **17**, 1–16.

McGETTRICK, A. F. & O'NEILL, L. A. (2004). The expanding family of MyD88-like adaptors in Toll-like receptor signal transduction. *Molecular Immunology* **41**, 577–582.

McGUIRE, W., HILL, A. V., ALLSOPP, C. E., GREENWOOD, B. M. & KWIATKOWSKI, D. (1994). Variation in the TNF-alpha promoter region associated with susceptibility to cerebral malaria. *Nature* **371**, 508–510.

MILLER, L. H., BARUCH, D. I., MARSH, K. & DOUMBO, O. K. (2002). The pathogenic basis of malaria. *Nature* **415**, 673–679.

MITAMURA, T., HANADA, K., KO-MITAMURA, E. P., NISHIJIMA, M. & HORII, T. (2000). Serum factors governing intraerythrocytic development and cell cycle progression of *Plasmodium falciparum*. *Parasitology International* **49**, 219–229.

MURAILLE, E., DE TREZ, C., BRAIT, M., DE BAETSELIER, P., LEO, O. & CARLIER, Y. (2003). Genetically resistant mice lacking MyD88-adapter protein display a high susceptibility to *Leishmania major* infection associated with a polarized Th2 response. *Journal of Immunology* **170**, 4237–4241.

NAGAI, Y., SHIMAZU, R., OGATA, H., AKASHI, S., SUDO, K., YAMASAKI, H., HAYASHI, S., IWAKURA, Y., KIMOTO, M. & MIYAKE, K. (2002). Requirement for MD-1 in cell surface expression of RP105/CD180 and B-cell responsiveness to lipopolysaccharide. *Blood* **99**, 1699–1705.

NAIK, R. S., BRANCH, O. H., WOODS, A. S., VIJAYKUMAR, M., PERKINS, D. J., NAHLEN, B. L., LAL, A. A., COTTER, R. J., COSTELLO, C. E., OCKENHOUSE, C. F., DAVIDSON, E. A. & GOWDA, D. C. (2000). Glycosylphosphatidylinositol anchors of *Plasmodium falciparum*: molecular characterization and naturally elicited antibody response that may provide immunity to malaria pathogenesis. *Journal of Experimental Medicine* **192**, 1563–1576.

O'NEILL, L. A. (2003). The role of MyD88-like adapters in Toll-like receptor signal transduction. *Biochemical Society Transactions* **31**, 643–647.

PFEIFFER, A., BOTTCHER, A., ORSO, E., KAPINSKY, M., NAGY, P., BODNAR, A., SPREITZER, I., LIEBISCH, G., DROBNIK, W., GEMPEL, K., HORN, M., HOLMER, S., HARTUNG, T., MULTHOFF, G., SCHUTZ, G., SCHINDLER, H., ULMER, A. J., HEINE, H., STELTER, F., SCHUTT, C., ROTHE, G., SZOLLOSI, J., DAMJANOVICH, S. & SCHMITZ, G. (2001). Lipopolysaccharide and ceramide docking to CD14 provokes ligand-specific receptor clustering in rafts. *European Journal of Immunology* **31**, 3153–3164.

PICHYANGKUL, S., YONGVANITCHIT, K., KUM-ARB, U., HEMMI, H., AKIRA, S., KRIEG, A. M., HEPPNER, D. G., STEWART, V. A., HASEGAWA, H., LOOAREESUWAN, S., SHANKS, G. D. & MILLER, R. S. (2004). Malaria blood stage parasites activate human plasmacytoid dendritic cells and murine dendritic cells through a Toll-like receptor 9-dependent pathway. *Journal of Immunology* **172**, 4926–4933.

POLTORAK, A., HE, X., SMIRNOVA, I., LIU, M. Y., VAN HUFFEL, C., DU, X., BIRDWELL, D., ALEJOS, E., SILVA, M., GALANOS, C., FREUDENBERG, M., RICCIARDI-CASTAGNOLI, P., LAYTON, B. & BEUTLER, B. (1998). Defective LPS signaling in C3H/HeJ and C57BL/10ScCr mice: mutations in Tlr4 gene. *Science* **282**, 2085–2088.

RINGWALD, P., PEYRON, F., VUILLEZ, J. P., TOUZE, J. E., LE BRAS, J. & DELORON, P. (1991). Levels of cytokines in plasma during *Plasmodium falciparum* malaria attacks. *Journal of Clinical Microbiology* **29**, 2076–2078.

ROPERT, C., ALMEIDA, I. C., CLOSEL, M., TRAVASSOS, L. R., FERGUSON, M. A., COHEN, P. & GAZZINELLI, R. T. (2001). Requirement of mitogen-activated protein kinases and I kappa B phosphorylation for induction of proinflammatory cytokines synthesis by macrophages indicates functional similarity of receptors triggered by glycosylphosphatidylinositol anchors from parasitic protozoa and bacterial lipopolysaccharide. *Journal of Immunology* **166**, 3423–3431.

ROPERT, C., FERREIRA, L. R., CAMPOS, M. A., PROCOPIO, D. O., TRAVASSOS, L. R., FERGUSON, M. A., REIS, L. F., TEIXEIRA, M. M., ALMEIDA, I. C. & GAZZINELLI, R. T. (2002). Macrophage signaling by glycosylphosphatidylinositol-anchored mucin-like glycoproteins derived from *Trypanosoma cruzi* trypomastigotes. *Microbes and Infection* **4**, 1015–1025.

SCANGA, C. A., ALIBERTI, J., JANKOVIC, D., TILLOY, F., BENNOUNA, S., DENKERS, E. Y., MEDZHITOV, R. & SHER, A. (2002). Cutting edge: MyD88 is required for resistance to *Toxoplasma gondii* infection and regulates parasite-induced IL-12 production by dendritic cells. *Journal of Immunology* **168**, 5997–6001.

SCHOFIELD, L., GEROLD, P., SCHWARZ, R. T. & TACHADO, S. (1994). Signal transduction in host cells mediated by glycosylphosphatidylinositols of the parasitic protozoa, or why do the parasitic protozoa have so many GPI molecules? *Brazilian Journal of Medical and Biological Research* **27**, 249–254.

SCHOFIELD, L. & HACKETT, F. (1993). Signal transduction in host cells by a glycosylphosphatidylinositol toxin of malaria parasites. *Journal of Experimental Medicine* **177**, 145–153.

SCHOFIELD, L., HEWITT, M. C., EVANS, K., SIOMOS, M. A. & SEEBERGER, P. H. (2002). Synthetic GPI as a candidate anti-toxic vaccine in a model of malaria. *Nature* **418**, 785–789.

SCHOFIELD, L., NOVAKOVIC, S., GEROLD, P., SCHWARZ, R. T., MCCONVILLE, M. J. & TACHADO, S. D. (1996). Glycosylphosphatidylinositol toxin of *Plasmodium* up-regulates intercellular adhesion molecule-1, vascular cell adhesion molecule-1, and E-selectin expression in vascular endothelial cells and increases leukocyte and parasite cytoadherence via tyrosine kinase-dependent signal transduction. *Journal of Immunology* **156**, 1886–1896.

SCHOFIELD, L., VILLAQUIRAN, J., FERREIRA, A., SCHELLEKENS, H., NUSSENZWEIG, R. & NUSSENZWEIG, V. (1987). Gamma interferon, CD8 + T cells and antibodies required for immunity to malaria sporozoites. *Nature* **330**, 664–666.

SCHOFIELD, L., VIVAS, L., HACKETT, F., GEROLD, P., SCHWARZ, R. T. & TACHADO, S. (1993). Neutralizing monoclonal antibodies to glycosylphosphatidylinositol, the dominant TNF-alpha-inducing toxin of *Plasmodium falciparum*: prospects for the immunotherapy of severe malaria. *Annals of Tropical Medicine and Parasitology* **87**, 617–626.

SCHROEDER, R. J., AHMED, S. N., ZHU, Y., LONDON, E. & BROWN, D. A. (1998). Cholesterol and sphingolipid enhance the Triton X-100 insolubility of glycosylphosphatidylinositol-anchored proteins by promoting the formation of detergent-insoluble ordered membrane domains. *Journal of Biological Chemistry* **273**, 1150–1157.

SHARMA, P., VARMA, R., SARASIJ, R. C., IRA, GOUSSET, K., KRISHNAMOORTHY, G., RAO, M. & MAYOR, S. (2004). Nanoscale organization of multiple GPI-anchored proteins in living cell membranes. *Cell* **116**, 577–589.

SILEGHEM, M., SAYA, R., GRAB, D. J. & NAESSENS, J. (2001). An accessory role for the diacylglycerol moiety of variable surface glycoprotein of African trypanosomes in the stimulation of bovine monocytes. *Veterinary Immunology Immunopathology* **78**, 325–339.

SIMONS, K. & TOOMRE, D. (2000). Lipid rafts and signal transduction. *Nature Reviews: Molecular Cell Biology* **1**, 31–39.

SUGUITAN, A. L., GOWDA, D. C., FOUDA, G., THUITA, L., ZHOU, A., DJOKAM, R., METENOU, S., LEKE, R. G. & TAYLOR, D. W. (2004). Lack of association between antibodies to *Plasmodium falciparum* glycosylphosphatidylinositols and malaria-associated placental changes in Cameroonian women with preterm and full-term deliveries. *Infection and Immunity* **72**, 5267–5273.

STEVENSON, M. M. & RILEY, E. M. (2004). Innate immunity to malaria. *Nature Reviews Immunology* **4**, 169–180.

TACHADO, S. D., GEROLD, P., MCCONVILLE, M. J., BALDWIN, T., QUILICI, D., SCHWARZ, R. T. & SCHOFIELD, L. (1996). Glycosylphosphatidylinositol toxin of *Plasmodium* induces nitric oxide synthase expression in macrophages and vascular endothelial cells by a protein tyrosine

kinase-dependent and protein kinase C-dependent signaling pathway. *Journal of Immunology* **156**, 1897–1907.

TACHADO, S. D., GEROLD, P., SCHWARZ, R., NOVAKOVIC, S., MCCONVILLE, M. & SCHOFIELD, L. (1997). Signal transduction in macrophages by glycosylphosphatidylinositols of *Plasmodium*, *Trypanosoma*, and *Leishmania*: activation of protein tyrosine kinases and protein kinase C by inositolglycan and diacylglycerol moieties. *Proceedings of the National Academy Sciences, USA* **94**, 4022–4027.

TACHADO, S. D. & SCHOFIELD, L. (1994). Glycosylphosphatidylinositol toxin of *Trypanosoma brucei* regulates IL-1 alpha and TNF-alpha expression in macrophages by protein tyrosine kinase mediated signal transduction. *Biochemistry and Biophysics Research Communications* **205**, 984–991.

TAKEDA, K., KAISHO, T. & AKIRA, S. (2003). Toll-like receptors. *Annual Review of Immunology* **21**, 335–376.

TAKEDA, K., TAKEUCHI, O. & AKIRA, S. (2002). Recognition of lipopeptides by Toll-like receptors. *Journal of Endotoxin Research* **8**, 459–463.

TRIANTAFILOU, K., TRIANTAFILOU, M. & DEDRICK, R. L. (2001 *a*). A CD14-independent LPS receptor cluster. *Nature Immunology* **2**, 338–345.

TRIANTAFILOU, K., TRIANTAFILOU, M., LADHA, S., MACKIE, A., DEDRICK, R. L., FERNANDEZ, N. & CHERRY, R. (2001 *b*). Fluorescence recovery after photobleaching reveals that LPS rapidly transfers from CD14 to hsp70 and hsp90 on the cell membrane. *Journal of Cell Science* **114**, 2535–2545.

TRIANTAFILOU, M., MIYAKE, K., GOLENBOCK, D. T. & TRIANTAFILOU, K. (2002). Mediators of innate immune recognition of bacteria concentrate in lipid rafts and facilitate lipopolysaccharide-induced cell activation. *Journal of Cell Science* **115**, 2603–2611.

TRIANTAFILOU, M. & TRIANTAFILOU, K. (2002). Lipopolysaccharide recognition: CD14, TLRs and the LPS-activation cluster. *Trends in Immunology* **23**, 301–304.

VARMA, R. & MAYOR, S. (1998). GPI-anchored proteins are organized in submicron domains at the cell surface. *Nature* **394**, 798–801.

VIJAYKUMAR, M., NAIK, R. S. & GOWDA, D. C. (2001). *Plasmodium falciparum* glycosylphosphatidylinositol-induced TNF-alpha secretion by macrophages is mediated without membrane insertion or endocytosis. *Journal of Biological Chemistry* **276**, 6909–6912.

WHITE, N. J. & HO, M. (1992). The pathophysiology of malaria. *Advances in Parasitology*. **31**, 83–173.

WRIGHT, S. D., RAMOS, R. A., TOBIAS, P. S., ULEVITCH, R. J. & MATHISON, J. C. (1990). CD14, a receptor for complexes of lipopolysaccharide (LPS) and LPS binding protein. *Science* **249**, 1431–1433.

YAMAMOTO, M., SATO, S., HEMMI, H., HOSHINO, K., KAISHO, T., SANJO, H., TAKEUCHI, O., SUGIYAMA, M., OKABE, M., TAKEDA, K. & AKIRA, S. (2003 *a*). Role of adaptor TRIF in the MyD88-independent toll-like receptor signaling pathway. *Science* **301**, 640–643.

YAMAMOTO, M., SATO, S., HEMMI, H., SANJO, H., UEMATSU, S., KAISHO, T., HOSHINO, K., TAKEUCHI, O., KOBAYASHI, M., FUJITA, T., TAKEDA, K. & AKIRA, S. (2002). Essential role for TIRAP in activation of the signalling cascade shared by TLR2 and TLR4. *Nature* **420**, 324–329.

YAMAMOTO, M., SATO, S., HEMMI, H., UEMATSU, S., HOSHINO, K., KAISHO, T., TAKEUCHI, O., TAKEDA, K. & AKIRA, S. (2003*b*). TRAM is specifically involved in the Toll-like receptor 4-mediated MyD88-independent signaling pathway. *Nature Immunology* **4**, 1144–1150.

ZHU, J., KRISHNEGOWDA, G. & GOWDA, D. C. (2005). Induction of proinflammatory responses in macrophages by the glycophosphatidylinositols (GPIs) of *Plasmodium falciparum. Journal of Biological Chemistry* **280**, 8617–8627.

Subversion of immune cell signal transduction pathways by the secreted filarial nematode product, ES-62

W. HARNETT[1]*, H. S. GOODRIDGE[2] and M. M. HARNETT[2]

[1] Department of Immunology, University of Strathclyde, Glasgow G4 0NR, UK
[2] Division of Immunology, Infection and Inflammation, University of Glasgow, G11 6NT, UK

SUMMARY

Filarial nematodes achieve longevity within the infected host by suppressing and modulating the host immune response. To do this, the worms actively secrete products that have been demonstrated to possess immunomodulatory properties. In this article we discuss the immunomodulatory effects of the phosphorylcholine-containing filarial nematode secreted glycoprotein ES-62. In particular we describe how it modulates intracellular signal transduction pathways in a number of different cells of the immune system, in particular B-lymphocytes, T-lymphocytes, macrophages and dendritic cells.

Key words: ES-62, filarial nematode, immune system cells, signal transduction.

INTRODUCTION

Filarial nematodes constitute highly successful parasites of vertebrates. Of the eight species known to infect humans, three – *Wuchereria bancrofti*, *Brugia malayi* and *Onchocerca volvulus* – are of major medical significance. Currently it is estimated that about 150 million people are infected with one or more of these worms and a significant proportion of these suffer severe health problems such as elephantiasis and blindness (WHO, 2000). Infection with filarial nematodes is typically life-long, with individual worms surviving for ∼10 years (Subramanian *et al.* 2004). Parasite longevity is due to suppression or modulation of the host immune system (reviewed in King, 2001 and Brattig, 2004) and there is increasing evidence that this is due to biologically active molecules secreted by the worms (reviewed in Maizels, Blaxter & Scott, 2001 and Harnett & Parkhouse, 1995). In this article we describe work in our laboratories over the last 12 years, which has examined immunomodulation by ES-62, a secreted product of the rodent filarial nematode *Acanthocheilonema viteae*. In particular, we focus on how the nematode product blocks key signal transduction pathways associated with immune system cell activation and polarisation.

ES-62, A PHOSPHORYLCHOLINE-CONTAINING GLYCOPROTEIN

ES-62 is only secreted by the post-infective lifecycle stages (L4 larvae and adult worms) of *A. viteae* and can be readily detected in the serum of infected jirds (Harnett *et al.* 1989; Stepek *et al.* 2002). In spite of this, the ES-62 gene is transcribed throughout the *A. viteae* lifecycle, although mRNA levels are considerably higher in adult worms than L3 larvae (∼5% adult levels) and microfilariae (<0·2% adult levels) (Stepek *et al.* 2004). ES-62 mRNA is translated into a glycoprotein of 62 kDa (including post-translational modifications) that has phosphorylcholine (PC) moieties attached via N-type glycans (reviewed in Harnett, Harnett & Byron, 2003; see Fig. 1) although the mature protein is secreted as a tetramer (Harnett *et al.* 1993, 2003; Ackerman *et al.* 2003). The number of PC-containing glycans present on each ES-62 molecule is currently unknown but the number of PC groups per glycan has been found to be variable. Many of ES-62's activities have been attributed to the presence of PC (Harnett & Harnett, 1999; Harnett *et al.* 1999).

ES-62 exerts its immunomodulatory effects on a variety of cells of the murine immune system including B and T lymphocytes as well as antigen presenting cells (APCs) such as dendritic cells (DCs) and macrophages (Harnett & Harnett, 1993; Harnett *et al.* 1998, 1999, Houston *et al.* 2000; Whelan *et al.* 2000; Goodridge *et al.* 2001; Wilson *et al.* 2003 *a, b*). Broadly, rather than acting in an immunosuppressive manner, the molecule induces a somewhat Th2/anti-inflammatory phenotype, characterized by the production of IL-10, with reduced levels of IL-12, IFN-γ and pro-inflammatory cytokines, and IgG1 rather than IgG2a antibodies. The signalling

* Corresponding author: Prof W. Harnett, Department of Immunology, University of Strathclyde, Glasgow G4 0NR, UK. Tel: 0141-548-3725. Fax: 0141-548-3427. E-mail: w.harnett@strath.ac.uk

A

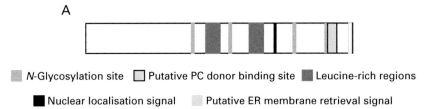

Fig. 1. ES-62 structural studies. The diagram depicts the location of key residues within the ES-62 sequence, including N-glycosylation sites, a possible site for interaction with PC donors, leucine-rich regions (likely to be involved in protein-protein interaction), and regions containing subcellular targeting motifs.

pathways targeted by ES-62 to achieve this immunomodulation are described below.

MODULATION OF INTRACELLULAR SIGNAL TRANSDUCTION

B cell signalling

The best understood mechanism of ES-62 action is its effect on B cells, which results in the suppression of proliferation following B cell antigen receptor (BCR) cross-linking. The BCR comprises a clonatypic antigen-binding component (surface immunoglobulin, sIg) and its accessory immunoreceptor tyrosine-based activatory motif (ITAM)-containing signal transducing molecules Igα and Igβ. Ligation of the BCR triggers protein tyrosine kinase (PTK) activity, resulting in tyrosine phosphorylation of the ITAMs (reviewed in Cambier, Pleiman & Clark, 1994 and Pleiman, D'Ambrosia and Cambier, 1994) and the recruitment of a number of key signal transducing pathways implicated in cellular activation and proliferation (Fig. 2). These include the phospholipase C (PLC)-γ, phosphoinositide-3-kinase (PI-3-K) and the Ras-Erk (extracellular-regulated kinase) mitogen-activated protein kinase (MAP kinase) signalling cascades.

Our studies have shown that ES-62 selectively targets these key signalling events following BCR ligation to disrupt the activation and proliferation of B cells. For example, it does not target the early BCR-coupled PLC-γ-mediated hydrolysis of phosphatidylinositol 4,5-bisphosphate (PIP$_2$), which generates the second messengers, inositol trisphosphate (IP$_3$) and diacylglycerol (DAG) that mobilise intracellular stores of calcium and activate protein kinase C isoforms respectively (Harnett & Harnett, 1993). Rather, ES-62 appears to selectively modulate the expression and activity of certain PKC isoforms in resting and BCR-stimulated B cells (Harnett & Harnett, 1993; Deehan, Harnett & Harnett, 1997). Thus, whilst ES-62 selectively down-regulates the expression of the α, β, ζ, δ and ι/λ isoforms, predominantly by stimulating proteolytic degradation, it up-regulates the expression of PKC-γ and PKC-ε in murine splenic B cells (Deehan et al. 1997). In addition, ES-62 acts to modulate PKC signalling

resulting from antigen receptor ligation of B cells by disrupting the normal activation and nuclear translocation patterns of the PKC-α and -ι/λ isoforms (Deehan et al. 1997). These data are consistent with proposals that PKC-α, -β and -ι/λ transduce key activation signals involved in the regulation of antigen-driven DNA synthesis and proliferation in B cells (BrickGhannam et al. 1994) such as phosphorylation of the nuclear protein lamin B and the induction and activation of NF-κB, Fos, Egr-1 and Myc (Hornbeck, Huang & Paul, 1988; Klemsz et al. 1989; Seyfert et al. 1990; Mittelstadt & DeFranco, 1993; Francis et al. 1995).

In addition to its effects on PKC signalling, pre-exposure to ES-62 selectively inhibits BCR-mediated recruitment of key proliferative pathways, such as the PI-3-K and Ras/Erk MAP kinase cascades (Deehan et al. 1998). An intriguing feature of these findings was that stimulation of B cells with ES-62 alone did not induce activation of Ras or PI-3-K despite the fact that ES-62 can induce activation of the PTKs Lyn and Syk (upstream regulators of Ras and PI-3-K) and Erk MAP kinase (downstream effector of Ras) (Deehan et al. 1998). This apparent discrepancy was resolved by our finding (Deehan et al. 1998; Deehan, Harnett & Harnett, 2001) that ES-62 does not mediate uncoupling of the BCR from the PI-3-K or Ras/Erk MAP kinase cascades by targeting activation of Syk or Lyn. Instead, it primes for the induction of the tyrosine phosphatase SHP-1, which negatively regulates activation via the BCR complex by dephosphorylating the Igα/β-ITAMs, thereby preventing recruitment of the Ras/Erk MAP kinase cascade. Moreover, ES-62 recruits additional negative regulatory elements of this pathway, namely RasGAP and the dual (thr/tyr) phosphatase Pac-1, to terminate any residual coupling of the BCR to Ras and Erk activity, respectively. This multi-pronged mechanism provides for a rapid and profound desensitisation of BCR-stimulated Erk MAP kinase signalling (Fig. 2).

ES-62 also modulates the activation of the two other major MAP kinase subfamilies, p38 and c-Jun N-terminal kinase (JNK) (Goodridge et al. 2005a). The precise targets of ES-62 in these pathways are unclear but our results, like those of others (Bagrodia et al. 1995), suggest that the BCR modulates p38 and

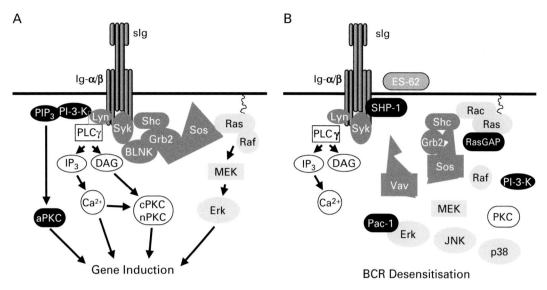

Fig. 2. BCR signalling in untreated (A) and ES-62-exposed (B) B cells. Following ligation of the B cell antigen receptor (BCR) of untreated B cells (A) the kinase, Lyn, tyrosine phosphorylates the Immunoreceptor Tyrosine Activation Motifs (ITAMs) on the accessory transducing molecules Ig-α and Ig-β resulting in the recruitment and activation of the PI-3-kinase (PI-3-K) and PLCγ-signalling pathways. Whilst PI-3-K activation results in the activation of atypical PKC isoforms (aPKC), PLCγ activation induces IP_3 and DAG generation, ultimately resulting in activation of classical (cPKC) and novel (nPKC) PKC isoforms. Binding of the adaptor proteins Shc and BLNK to the phosphorylated ITAMs leads to the recruitment of the Grb2Sos complexes (Grb2 is an adaptor protein which binds Sos, a guanine nucleotide exchange factor) required for activation of the GTPase, Ras. Active Ras initiates the Erk mitogen-activated protein (MAP) kinase cascade by binding and activating the ser/thr kinase, Raf leading to stimulation of the thr/tyr kinase MEK and consequent activation and nuclear translocation of the ser/thr kinase Erk. ES-62/PC signalling (B) disrupts BCR coupling to the PI-3-K cascade as well as targeting major negative regulatory sites in the control of the Erk, p38 and JNK MAP kinase cascades. Firstly, ES-62 signalling promotes the BCR-activation of SHP-1 tyrosine phosphatase to prevent initiation of BCR signalling by maintaining the ITAMs in a resting, dephosphorylated state and hence prevents recruitment of the ShcGrb2Sos complexes required to activate the Ras- and Rac-MAP kinase cascades. Secondly, ES-62 signalling promotes the BCR-mediated recruitment of RasGAP to terminate ongoing Ras signals. In addition, ES-62 is also likely to target MAP kinase activation by downregulating PKC isoform expression. Finally, ES-62-signalling promotes the BCR-driven association of the nuclear MAP kinase dual (thr/tyr) phosphatase, Pac-1 with Erk to terminate any ongoing Erk signals. This multi-pronged mechanism results in a rapid and profound desensitisation of BCR coupling to the MAP kinase cascades.

JNK signalling in a Vav- and Rac-dependent manner (Goodridge *et al.* 2005*a*). The finding that ES-62 alone can selectively stimulate Syk, Lyn and MAP kinase activation, whilst it does not appear to modulate Ras, Rac or PI-3-K activity (Deehan *et al.* 1997, 1998), also suggests that MAP kinases can be activated in B cells via alternative PTK-dependent pathways. This proposal is consistent with the increasing evidence for Ras-independent pathways of Erk MAP kinase activation involving lipid second messengers and PKC (Buscher *et al.* 1995). We have not, however, found any evidence to support the proposal that ES-62 or BCR stimulates Erk MAP kinase activity via lipid second messengers derived from phosphatidylcholine-specific phospholipase C (PtdCho-PLC), PtdCho-phospholipase D (PtdCho-PLD) or sphingomyelinase-dependent pathways (Deehan *et al.* 1998). However, PKC-α has been reported to mediate activation of Erk MAP kinase via Raf and MAP kinase kinase (MEK) pathways (Berra *et al.* 1995; Pelech, 1996) and this is consistent not only with reports that PKC activity plays a role in

coupling the BCR to ErkMAP kinases (Campbell, 1999), but also provides a rationale for our finding that, whilst prolonged pre-treatment with ES-62 acts to reduce this PKC activity, ES-62 initially up-regulates PKC-α expression (Harnett & Harnett, 1993; Deehan *et al.* 1997).

T cell signalling

Although ES-62 exerts little or no direct effects on antigen-/mitogen-driven proliferation or cytokine secretion in T cells, in a manner analogous to that observed for the suppression of antigen receptor stimulation of B cells, the nematode product suppresses anti-CD3-induced proliferation of Jurkat T cells and also promotes concanavalin A-induced growth arrest of these cells (Harnett *et al.* 1998). The precise sequence of signalling events underlying these effects has not been fully elucidated, but it is clear that ES-62-mediated desensitisation of TCR signalling is associated with disruption of TCR coupling to PLD, PKC, PI-3-K and Ras-Erk MAP kinase signalling but, as with B cells, not the PLC-mediated generation of inositol phosphates (Harnett *et al.* 1998). Again, PC appears to be the active moiety, since culture with PC or PC-BSA has comparable effects to ES-62 on the coupling of the TCR to PTK activation (ZAP-70, Lck and Fyn recruitment) and the Ras-Erk MAP kinase signalling cascades (Harnett *et al.* 1998, 1999). These findings are consistent with an earlier report showing that PC-containing molecules of the human filarial nematode, *B. malayi*, inhibit the response of human T cells to mitogens (Lal *et al.* 1990).

Macrophage and DC signalling

Induction of cytokine production by macrophages and DCs in response to many pathogen products occurs following ligation of a family of recently identified pattern recognition receptors, known as Toll-like receptors (TLRs; reviewed in Takeda, Kaisho & Akira, 2003). TLRs are thought to recognise specific molecular motifs of host as well as pathogen origin, including pathogen-associated molecular patterns (PAMPs). For example, TLR4 is required for the detection of and response to bacterial LPS whereas BLP and CpG DNA motifs are recognized by TLR2 and TLR9 respectively. TLR signals are transduced via adaptor molecules including MyD88 (reviewed in O'Neill, Fitzgerald & Bowie, 2003 and Akira & Takeda, 2004) resulting in the activation of various signalling pathways including the MAP kinase cascades and NF-κB.

Treatment of macrophages with ES-62 induces tyrosine phosphorylation of a number of proteins (Goodridge *et al.* 2005*a*) and modulates the activation of members of all three major MAP kinase subfamilies (Erk, p38 and JNK), which are involved in the regulation of cytokine production (Goodridge *et al.* 2001, 2003). For example, suppression of IL-12 by ES-62 is likely to be due in part to Erk MAP kinase-mediated suppression of the p40 subunit of this heterodimeric cytokine, since ES-62 inhibition can be rescued by pre-treatment with the Erk MAP kinase kinase (MEK-1) inhibitor PD98059 (Goodridge *et al.* 2003). ES-62 treatment also suppresses the activation of the p38 (Goodridge *et al.* 2003) and JNK MAP kinases (Goodridge *et al.* 2005*a*), which are required for the production of IL-12 as well as IL-6 and TNF-α, suggesting another mechanism whereby suppression of these cytokines is likely to be achieved. Furthermore, preliminary analysis indicates that ES-62 regulates gene induction by modulating the activation and gene promoter region binding of the transcription factors NF-κB and IRF-1 (Goodridge *et al.* 2005*a*).

We have recently shown that ES-62 achieves its modulation of macrophage and DC activation in a TLR4-dependent manner (Goodridge *et al.* 2005*b*). Low-level cytokine induction by ES-62 alone was abolished in macrophages/DCs from TLR4 ko mice and also MyD88 ko mice. Similarly ES-62-mediated suppression of cytokine induction by TLR ligands (BLP and CpG) was dependent on the presence of TLR4. In contrast ES-62 effects were TLR2- and TLR-6 independent. Modulation of surface expression of MHC II and costimulatory molecules (CD40, CD80, CD86) was also suppressed in DCs from TLR4 ko mice, although these latter effects were only partially dependent on the presence of MyD88. Interestingly macrophages and DCs from C3H/HeJ mice, which are unresponsive to LPS due to their production of a defective form of TLR4 resulting from a point mutation in the intracellular TIR domain, remain responsive to ES-62 as evidenced by the modulation of cytokine production and costimulatory molecule expression. Thus it appears that TLR4 must be present but not necessarily fully functional for ES-62 responsiveness. This may be due to 'non-classical' coupling to downstream signal transduction pathways by TLR4 or the recruitment of a signalling co-receptor.

It is not currently clear whether TLR4 is required for the direct recognition of ES-62. Preliminary evidence suggests that the PC moiety of ES-62 is at least in part required for its recognition (unpublished data) and hence it is tempting to speculate that TLR4 may recognize this molecular pattern, especially since other PC-containing molecules have been reported to act via TLR4 (Miller *et al.* 2003; Walton *et al.* 2003*a, b*).

CONCLUDING REMARKS

Our studies over the past twelve years have illustrated how a single parasite glycoprotein can cause profound modulation of the host immune response via co-ordinated targeting of signal transduction pathways in multiple cells of the immune system. Importantly, ES-62 is not unique to *A. viteae*; homologues of ES-62 are also found in other filarial nematodes, including the human parasites *B. malayi* and *O. volvulus* (Nor *et al.* 1997; Stepek *et al.* 2002, 2004). For example, *A. viteae* ES-62 shares 77·1% homology with an ES-62 cDNA of *B. malayi* and the latter contains the important post-translational modification of PC attached to N-type glycans (Goodridge *et al.* 2005*c*). Thus the work we are describing is truly of relevance to human filariasis. In addition, PC-containing molecules are produced by a diverse range of pathogens other than filarial nematodes (reviewed in Harnett & Harnett, 1999). What is currently unknown is what effects these molecules have on immune cell signalling but it will be of interest to determine whether they subvert it in a manner analogous to ES-62.

ACKNOWLEDGEMENTS

Funding was from the Wellcome Trust, the BBSRC, the Leverhulme Trust and the MRC.

REFERENCES

ACKERMAN, C. J., HARNETT, M. M., HARNETT, W., KELLY, S. M., SVERGUN, D. I. & BYRON, O. (2003). 19 A solution structure of the filarial nematode immunomodulatory protein, ES-62. *Biophysical Journal* **84**, 489–500.

AKIRA, S. & TAKEDA, K. (2004). Toll-like receptor signalling. *Nature Reviews Immunology* **4**, 499–511.

BAGRODIA, S., DERIJARD, B., DAVIS, R. J. & CERIONE, R. A. (1995). Cdc42 and Pak-mediated signaling leads to Jun kinase and P38 mitogen-activated protein-kinase activation. *Journal of Biological Chemistry* **270**, 27995–27998.

BERRA, E., DIAZ-MECO, M. T., LOZANO, J., FRUTOS, S., MUNICIO, M. M., SANCHEZ, P., SANZ, L. & MOSCAT, J. (1995). Evidence for a role of MEK and MAPK during signal transduction by protein kinase C ξ. *EMBO Journal* **14**, 6157–6163.

BRATTIG, N. W. (2004). Pathogenesis and host responses in human onchocerciasis: impact of *Onchocerca* filariae and *Wolbachia* endobacteria. *Microbes and Infection* **6**, 113–128.

BRICKGHANNAM, C., ERICSON, M. L., SCHELLE, I. & CHARRON, D. (1994). Differential regulation of mRNAs encoding protein kinase C isoenzymes in activated human B cells. *Human Immunology* **41**, 216–224.

BUSCHER, D., HIPSKIND, R. A., KRAUTWALD, S., REIMANN, T. & BACCARINI, M. (1995). Ras-dependent and -independent pathways target the mitogen-activated protein kinase network in macrophages. *Molecular Cell Biology* **15**, 466–475.

CAMBIER, J. C., PLEIMAN, C. & CLARK, M. R. (1994). Signal transduction by the B cell antigen receptor and its coreceptors. *Annual Review of Immunology* **12**, 457–486.

CAMPBELL, K. S. (1999). Signal transduction from the B cell antigen receptor. *Current Opinion in Immunology* **11**, 256–264.

DEEHAN, M., HARNETT, M. & HARNETT, W. (1997). A filarial nematode secreted product differentially modulates expression and activation of protein kinase C isoforms in B lymphocytes. *Journal of Immunology* **159**, 6105–6111.

DEEHAN, M. R., FRAME, M. J., PARKHOUSE, R. M., SEATTER, S. D., REID, S. D., HARNETT, M. M. & HARNETT, W. (1998). A phosphorylcholine-containing filarial nematode-secreted product disrupts B lymphocyte activation by targeting key proliferative signaling pathways. *Journal of Immunology* **160**, 2692–2699.

DEEHAN, M. R., HARNETT, W. & HARNETT, M. M. (2001). A filarial nematode-secreted phosphorylcholine-containing glycoprotein uncouples the B cell antigen receptor from extracellular signal-regulated kinase-mitogen-activated protein kinase by promoting the surface Ig-mediated recruitment of Src homology 2 domain-containing tyrosine phosphatase-1 and Pac-1 mitogen-activated kinase-phosphatase. *Journal of Immunology* **166**, 7462–7468.

FRANCIS, D. A., KARRAS, J. G., KE, X. Y., SEN, R. & ROTHSTEIN, T. L. (1995). Induction of the transcription factors NF-kappaB, AP-1 and NF-AT during B cell stimulation through the CD40 receptor. *International Immunology* **7**, 151–161.

GOODRIDGE, H. S., DEEHAN, M. R., HARNETT, W. & HARNETT, M. M. (2005*a*). Subversion of immunological signalling by a filarial nematode phosphorylcholine-containing secreted product. *Cell Signalling*, **17**, 11–16.

GOODRIDGE, H. S., HARNETT, W., LIEW, F. Y. & HARNETT, M. M. (2003). Differential regulation of interleukin-12 p40 and p35 induction via Erk mitogen-activated protein kinase-dependent and -independent mechanisms and the implications for bioactive IL-12 and IL-23 responses. *Immunology*, **109**, 415–425.

GOODRIDGE, H. S., MARSHALL, F. A., ELSE, K. J., HOUSTON, K. M., EGAN, C., AL-RIYAMI, L., LIEW, F. Y., HARNETT, W. & HARNETT, M. M. (2005*b*). Immunomodulation via novel use of TLR4 by the filarial nematode phosphorylcholine-containing secreted product, ES-62. *Journal of Immunology* **174**, 284–293.

GOODRIDGE, H. S., STEPEK, G., HARNETT, W. & HARNETT, M. M. (2005*c*). Signalling mechanisms underlying subversion of the immune response by a filarial nematode secreted product. *Immunology*, In press.

GOODRIDGE, H. S., WILSON, E. H., HARNETT, W., CAMPBELL, C. C., HARNETT, M. M. & LIEW, F. Y. (2001). Modulation of macrophage cytokine production by ES-62, a secreted product of the filarial nematode *Acanthocheilonema viteae*. *Journal of Immunology* **167**, 940–945.

HARNETT, M. M., DEEHAN, M. R., WILLIAMS, D. M. & HARNETT, W. (1998). Induction of signalling anergy via the T-cell receptor in cultured Jurkat T cells by pre-exposure to a filarial nematode secreted product. *Parasite Immunology*, **20**, 551–563.

HARNETT, W., DEEHAN, M. D., HOUSTON, K. M. & HARNETT, M. M. (1999). Immunomodulatory properties of a phosphorylcholine-containing secreted filarial glycoprotein. *Parasite Immunology* **21**, 601–608.

HARNETT, W., GRAINGER, M., KAPIL, A., WORMS, M. J. & PARKHOUSE, R. M. E. (1989). Origin, kinetics of circulation and fate *in vivo* of the major excretory-secretory product of *Acanthocheilonema viteae*. *Parasitology Today* **99**, 229–239.

HARNETT, W. & HARNETT, M. M. (1993). Inhibition of murine B cell proliferation and down-regulation of protein kinase C levels by a phosphorylcholine-containing filarial excretory-secretory product. *Journal of Immunology* **151**, 4829–4837.

HARNETT, W. & HARNETT, M. M. (1999). Phosphorylcholine: Friend or foe of the immune system? *Immunology Today* **20**, 125–129.

HARNETT, W., HARNETT, M. M. & BYRON, O. (2003). Structural/functional aspects of ES-62 – a secreted immunomodulatory phosphorylcholine-containing filarial nematode glycoprotein. *Current Protein and Peptide Science* **4**, 59–72.

HARNETT, W., HOUSTON, K. M., AMESS, R. & WORMS, M. J. (1993). *Acanthocheilonema viteae*: phosphorylcholine is attached to the major excretory-secretory product via an N-linked glycan. *Experimental Parasitology* **77**, 498–502.

HARNETT, W. & PARKHOUSE, R. M. E. (Eds.) (1995). *Structure and Function of Nematode Surface and Excretory-Secretory Products*, M/S Narendra Publication House, New Delhi.

HORNBECK, P., HUANG, K. P. & PAUL, W. E. (1988). Lamin B is rapidly phosphorylated in lymphocytes after activation of protein kinase C. *Proceedings of the National Academy of Sciences, USA* **85**, 2279–2283.

HOUSTON, K., WILSON, E. H., EYRES, L., BROMBACHER, F., HARNETT, M. M., ALEXANDER, J. & HARNETT, W. (2000). The presence of phosphorylcholine on a filarial nematode protein influences the IgG subclass response to a molecule and by a mechanism dependent on IL-10. *Infection and Immunity* **68**, 5466–5468.

KING, C. L. (2001). Transmission intensity and human immune responses to lymphatic filariasis. *Parasite Immunology* **23**, 363–371.

KLEMSZ, M. J., JUSTEMENT, L. B., PALMER, E. & CAMBIER, J. C. (1989). Induction of c-fos and c-myc expression during B cell activation by IL-4 and immunoglobulin binding ligands. *Journal of Immunology* **143**, 1032–1039.

LAL, R. B., KUMARASWAMI, V., STEEL, C. & NUTMAN, T. B. (1990). Phosphorylcholine-containing antigens of *Brugia malayi* non-specifically suppress lymphocyte function. *American Journal of Tropical Medicine and Hygiene* **42**, 56–64.

MAIZELS, R. M., BLAXTER, M. L. & SCOTT, A. L. (2001). Immunological genomics of *Brugia malayi*: filarial genes implicated in immune evasion and protective immunity. *Parasite Immunology* **23**, 327–344.

MILLER, Y. I., VIRIYAKOSOL, S., BINDER, C. J., FERAMISCO, J. R., KIRKLAND, T. N. & WITZTUM, J. L. (2003). Minimally modified LDL binds to CD14, induces macrophage spreading via TLR4/MD-2, and inhibits phagocytosis of apoptotic cells. *Journal of Biological Chemistry* **278**, 1561–1568.

MITTELSTADT, P. R. & DEFRANCO, A. L. (1993). Induction of early response genes by crosslinking membrane Ig on B lymphocytes. *Journal of Immunology* **150**, 4822–4832.

NOR, Z. M., HOUSTON, K. M., DEVANEY, E. & HARNETT, W. (1997). Variation in the nature of attachment of phosphorylcholine to excretory-secretory products of adult *Brugia pahangi*. *Parasitology Today* **114**, 257–262.

O'NEILL, L. A., FITZGERALD, K. A. & BOWIE, A. G. (2003). The Toll-IL-1 receptor adaptor family grows to five members. *Trends in Immunology* **24**, 286–290.

PELECH, S. L. (1996). Kinase connections on the cellular internet. *Current Biology* **6**, 551–554.

PLEIMAN, C. M., D'AMBROSIA, D. & CAMBIER, J. C. (1994). The B-cell antigen receptor complex: Structure and signal transduction. *Immunology Today* **15**, 393–398.

SEYFERT, V. L., McMAHON, S., GLENN, W., CAO, X., SUKHATME, V. P. & MONROE, J. G. (1990). Egr-1 expression in surface Ig-mediated B cell activation. *Journal of Immunology* **145**, 3647–3653.

STEPEK, G., AUCHIE, M., TATE, R., WATSON, K., RUSSELL, D. G., DEVANEY, E. & HARNETT, W. (2002). Expression of the filarial nematode phosphorylcholine-containing glycoprotein, ES-62, is stage-specific. *Parasitology Today* **125**, 155–164.

STEPEK, G., HOUSTON, K. M., GOODRIDGE, H. S., DEVANEY, E. & HARNETT, W. (2004). Stage-specific and species-specific differences in the production of the mRNA and protein for the filarial nematode secreted product, ES-62. *Parasitology Today* **128**, 91–98.

SUBRAMANIAN, S., STOLK, W. A., RAMAIAH, K. D., PLAISIER, A. P., KRISHNAMOORTHY, K., VAN OORTMARSSEN, G. J., DOMINIC AMALRAJ, D., HABBEMA, J. D. & DAS, P. K. (2004). The dynamics of *Wuchereria bancrofti* infection: a model-based analysis of longitudinal data from Pondicherry, India. *Parasitology* **128**, 467–482.

TAKEDA, K., KAISHO, T. & AKIRA, S. (2003). Toll-like receptors. *Annual Review of Immunology* **21**, 335–376.

WALTON, K. A., COLE, A. L., YEH, M., SUBBANAGOUNDER, G., KRUTZIK, S. R., MODLIN, R. L., LUCAS, R. M., NAKAI, J., SMART, E. J., VORA, D. K. & BERLINER, J. A. (2003a). Specific phospholipid oxidation products inhibit ligand activation of toll-like receptors 4 and 2. *Arteriosclerosis, Thrombosis and Vascular Biology* **23**, 1197–1203.

WALTON, K. A., HSIEH, X., GHARAVI, N., WANG, S., WANG, G., YEH, M., COLE, A. L. & BERLINER, J. A. (2003b). Receptors involved in the oxidized 1-palmitoyl-2-arachidonoyl-sn-glycero-3-phosphorylcholine-mediated synthesis of interleukin-8. A role for Toll-like receptor 4 and a glycosylphosphatidylinositol-anchored protein. *Journal of Biological Chemistry* **278**, 29661–29666.

WHELAN, M., HARNETT, M. M., HOUSTON, K. M., PATEL, V., HARNETT, W. & RIGLEY, K. P. (2000). A filarial nematode-secreted product signals dendritic cells to acquire a phenotype that drives development of Th2 Cells. *Journal of Immunology* **164**, 6453–6460.

WHO (2000). Filariasis. World Health Organisation, Geneva.

WILSON, E. H., DEEHAN, M. R., KATZ, E., BROWN, K. S., HOUSTON, K. M., O'GRADY, J., HARNETT, M. M. & HARNETT, W. (2003a). Hyporesponsiveness of murine B lymphocytes exposed to the filarial nematode secreted product ES-62 *in vivo*. *Immunology* **109**, 238–245.

WILSON, E. H., KATZ, E., GOODRIDGE, H. S., HARNETT, M. & HARNETT, W. (2003b). *In vivo* activation of murine peritoneal B1 cells by the filarial nematode phosphorylcholine-containing glycoprotein ES-62. *Parasite Immunology* **25**, 463–466.

Printed in the United States
by Baker & Taylor Publisher Services